Ma... W9-BDJ-448

PREFACE

A conference on Vector Space Measures and Applications was held at Trinity College, University of Dublin, during the week June 26 to July 2, 1977. Over one hundred and twenty mathematicians from eighteen countries participated. More than seventy five lectures were given, the texts of many of these appearing in the Proceedings.

The original intention of the Conference organisers was to arrange a fairly narrow range of featured topics. However, as the Conference planning progressed, it became clear that there was a great deal of interest in vector space measure theory by mathematicians, working in a much broader spectrum of fields who saw connections between current research in vector space measures and their own fields of research. Consequently, there were sessions on probability theory, distribution theory, quantum field theory, vector measures, functional analysis and real and complex analysis in infinite dimensions.

With the exception of twenty papers on real and complex analysis in infinite dimensions, which will be published separately, these Proceedings (in two volumes) contain the written and expanded texts of most of the papers given at the Conference.

The organising Committee consisted of Richard M. Aron (Trinity College Dublin), Paul Berner (Trinity College Dublin), Philip Boland (University College Dublin), Seán Dineen (University College Dublin), John Lewis (The Dublin Institute for Advanced Studies) and Paul McGill (The New University of Ulster, Coleraine). The Conference was made possible through the interest, cooperation and financial support of the European Research Office as well as Trinity College Dublin, University College Dublin, The Royal Irish Academy, The Dublin Institute for Advanced Studies, The Bank of Ireland and Bord Failte.

Richard M. Aron,
School of Mathematics,
Trinity College Dublin,
Dublin 2, Ireland.

Seán Dineen,
Department of Mathematics,
University College Dublin,
Belfield, Dublin 4, Ireland.

Lecture Notes in Mathematics

Edited by A. Dold and B. Eckmann

645

Vector Space Measures and Applications II

Proceedings, Dublin 1977

Edited by
R. M. Aron and S. Dineen

Springer-Verlag
Berlin Heidelberg New York 1978

Editors
Richard M. Aron
School of Mathematics
39 Trinity College
Dublin 2, Ireland

Seán Dineen
Department of Mathematics
University College Dublin
Belfield
Dublin 4, Ireland

AMS Subject Classifications (1970): 28-XX, 35-XX, 46-XX, 58-XX, 60-XX, 81-XX

ISBN 3-540-08669-2 Springer-Verlag Berlin Heidelberg New York
ISBN 0-387-08669-2 Springer-Verlag New York Heidelberg Berlin

This work is subject to copyright. All rights are reserved, whether the whole or part of the material is concerned, specifically those of translation, re-printing, re-use of illustrations, broadcasting, reproduction by photocopying machine or similar means, and storage in data banks. Under § 54 of the German Copyright Law where copies are made for other than private use, a fee is payable to the publisher, the amount of the fee to be determined by agreement with the publisher.

© by Springer-Verlag Berlin Heidelberg 1978
Printed in Germany

Printing and binding: Beltz Offsetdruck, Hemsbach/Bergstr.
2141/3140-543210

C O N T E N T S

CONTENTS OF VOLUME ONE *

* Volume I appeared as volume 644 in Lecture Notes in Mathematics

CONVERGENCE PRESQUE PARTOUT DES SUITES DE FONCTIONS MESURABLES ET APPLICATIONS

par D. BUCCHIONI et A. GOLDMAN

Dept. de Mathematiques,
Univ. Claude Bernard (Lyon I),
69 Villeurbanne, Lyon, France.

INTRODUCTION.

Nous avons annoncé dans (1) et (2) des résultats concernant la structure des suites de fonctions numériques mesurables dont aucune sous-suite ne converge presque partout. Dans le présent papier, nous redonnons de façon succinte les points essentiels de (2) complétés par diverses applications issues d'un travail de (4). On étudie notamment la mesurabilité des fonctions vectorielles à valeurs dans l'espace $C_S(K)$ des fonctions continues sur un compact K, cet espace étant muni de la topologie de la convergence simple sur K ; on en déduit aussi quelques propriétés des mesures de Baire sur un tel espace. Dans toute la suite, on désignera par (Ω,Σ,μ) un espace mesuré abstrait, la mesure μ étant toujours supposée positive et bornée.

1. UN CRITERE DE NON-MESURABILITE.

Nous donnons ici un critère pour qu'une application $f : (\Omega,\Sigma,\mu) \to R$ soit non-mesurable ; ce critère, dont l'énoncé est un peu technique, est en fait essentiel pour la compréhension de ce qui suit.

(1.1) THEOREME.
Soit (Ω,Σ,μ) *un espace mesuré complet et soit* $f : \Omega \to R$ *une fonction numérique non mesurable relativement à la tribu* Σ *et à la tribu borélienne* $\mathcal{B}$ *de* R. *Alors la propriété suivante est réalisée :*

(P) $\begin{cases} \textit{Il existe } Y \in \Sigma,\ \mu(Y) > 0 \textit{ et deux nombres } r \in R,\ \delta > 0 \textit{ tels que pour} \\ \textit{tout } A \subset Y,\ A \in \Sigma,\ \mu(A) > 0, \textit{ on puisse trouver } x \textit{ et } y \in A \textit{ vérifiant} \\ f(x) > r+\delta \textit{ et } f(y) < r. \end{cases}$

PREUVE.
Nous avons donné dans (2) une preuve basée sur l'axiome du choix. Notons à ce sujet qu'il est possible de donner une démonstration classique, mais plus longue, par récurrence dénombrable.

Que la tribu Σ soit complète ou non le critère (P) est une condition suffisante pour qu'une fonction numérique f définie sur l'espace mesuré (Ω,Σ,μ) soit non mesurable.

On a donc

(1.2) THEOREME.

Pour qu'une fonction numérique f *définie sur un espace mesuré quelconque* (Ω,Σ,μ) *soit non mesurable (relativement à la tribu borélienne* $\mathcal{B}$ *de* R*), il suffit qu'elle vérifie la condition (P) du théorème (1.1).*

Dans le cas où μ est une mesure de Radon sur un espace topologique T, on a le résultat plus particulier suivant :

(1.3) COROLLAIRE.

Soit μ *une mesure de Radon sur un espace topologique* T. *Pour qu'une application* $f : T \to R$ *ne soit pas* μ*-mesurable il faut et il suffit qu'elle vérifie le critère suivant :*

(P_R) *Il existe un compact* K_o *de* T, $\mu(K_o) > 0$, *et deux nombres* $r \in R$, $\delta > 0$ *tels que, pour tout compact* $K \subset K_o$, $\mu(K) > 0$, *on puisse trouver* x *et* $y \in K$ *vérifiant* $f(x) > r+\delta$ *et* $f(y) < r$.

Nous allons maintenant montrer que si une suite (f_n) de fonctions numériques mesurables n'a aucune sous-suite qui converge presque partout, alors elle "se comporte globalement" comme une fonction non mesurable ; ceci constituera l'essentiel du paragraphe 2.

2. SUITES DE FONCTIONS NUMERIQUES MESURABLES DONT AUCUNE SOUS-SUITE NE CONVERGE PRESQUE PARTOUT.

Le résultat principal est résumé par le théorème suivant :

(2.1) THEOREME.

Soit (f_n) *une suite de fonctions mesurables. Alors :*

a) Pour toute partie infinie M *de* N *et pour tous nombres* $r \in R$, $\delta > 0$, *l'ensemble* K_M *des points* $x \in \Omega$ *pour lesquels il existe deux parties infinies* P *et* P' *de* M *telles que* $f_n(x) > r+\delta$ *pour tout* $n \in P$ *et* $f_n(x) < r$ *pour tout* $n \in P'$, *appartient à la tribu complétée* $\hat{\Sigma}$.

b) Si de plus aucune sous-suite de (f_n) *ne converge presque partout, il existe une partie infinie* M *de* N *et deux nombres* $r \in R$, $\delta > 0$ *tels que, pour toute partie infinie* $L \overset{p.s.}{\subset} M$ *(i.e.* $L \setminus M$ *est fini) on ait :*

$$\mu^*(K_M) = \mu^*(K_L) > 0.$$

PREUVE.

Le point a) est établi dans (2) ; pour obtenir le point b), on commence par démontrer l'existence d'une partie infinie M' de N et de deux nombres $r \in R$, $\delta > 0$ tels que pour toute partie $L \underset{p.s.}{\subset} M'$, l'ensemble K_L soit de mesure extérieure $\mu^*(K_L)$ non nulle. Le couple (r,δ) étant maintenant fixé, on construit alors, par une récurrence transfinie, la partie infinie M de N souhaitée.

Le théorème (2.1) permet alors d'obtenir :

(2.2) COROLLAIRE.

Soit (f_n) *une suite de fonctions mesurables dont aucune sous-suite ne converge presque partout. Il existe* $Y \in \Sigma$, $\mu(Y) > 0$, *deux nombres* $r \in R$, $\delta > 0$ *et une partie infinie* M *de* N *tels que, pour toute partie infinie* $L \underset{p.s.}{\subset} M$ *et pour tout* $A \in \Sigma$, $A \subset Y$, $\mu(A) > 0$, *on puisse trouver* x *et* $y \in A$ *vérifiant* $f_n(x) > r+\delta$ *et* $f_n(y) < r$ *pour une infinité d'indices* $n \in L$.

Supposons maintenant que la tribu Σ vérifie la condition card $\overline{\Sigma} \leqslant 2^{\aleph_0}$ où $\overline{\Sigma}$ désigne l'ensemble des parties $A \in \Sigma$ de mesure non nulle (c'est par exemple le cas lorsque la tribu Σ est dénombrablement engendrée). Du théorème (2.2) on peut aussi déduire :

(2.4) THEOREME.

Soit (f_n) *une suite simplement bornée de fonctions mesurables dont aucune sous-suite ne converge presque partout. Alors, avec l'hypothèse du continu, il existe une fonction non mesurable* f *qui est valeur d'adhérence de la suite* (f_n) *pour la topologie de la convergence simple.*

PREUVE.

Elle s'obtient par un procédé de construction ordinale ; pour plus de détails on pourra se référer à (2).

REMARQUE.

D.H. FREMLIN a démontré dans (3), et par une méthode totalement différente, le résultat suivant :

THEOREME (FREMLIN).

Soit (Ω,Σ,μ) *un espace mesuré parfait et soit* (f_n) *une suite de fonctions mesurables. Alors l'une des deux assertions suivantes est réalisée :*

a) Il existe une sous-suite (f_{n_k}) *qui converge presque partout.*

b) Il existe une sous-suite (f_{n_k}) *n'ayant aucune valeur d'adhérence (pour la topologie de la convergence simple) mesurable.*

On peut établir (voir par exemple (2)), que supposer l'espace mesuré (Ω,Σ,μ) parfait revient en fait à supposer que μ est une mesure de Radon sur l'espace R^N et ainsi la démonstration de Fremlin (basée sur les propriétés de la mesure de Haar) ne peut pas s'appliquer dans le cadre abstrait. D'un autre côté, notre méthode ne permet pas de retrouver entièrement le résultat de (3).

3. APPLICATIONS A L'ETUDE DES ESPACES DE BANACH NE CONTENANT AUCUN SOUS-ESPACE ISOMORPHE A $\ell^1(N)$.

Dans (5), R. HAYDON a obtenu des conditions nécessaires et suffisantes, s'exprimant en termes de mesurabilité, pour qu'un espace de Banach contienne (isomorphiquement) l'espace $\ell^1(N)$. Par une technique analogue à celle utilisée par H.P. ROSENTHAL dans (8) pour l'étude des fonctions de la première classe de Baire, on peut déduire du théorème (1.1) le lemme suivant :

(3.1) LEMME.

Soient K *un espace compact,* μ *une mesure de Radon sur* K *et soit* A *une famille uniformément bornée de fonctions continues sur* K. *S'il existe une fonction non* μ*-mesurable* f *qui est valeur d'adhérence de* A *dans l'espace* R^K, *alors* A *contient une suite* (f_n) *équivalente, pour la norme de l'espace* $L^\infty(K,\mu)$, *à une base de l'espace* $\ell^1(N)$.

Ce lemme permet alors de retrouver (et sans utiliser la convexité) le résultat de R. HAYDON évoqué ci-dessus :

(3.2) THEOREME (HAYDON).

Pour un espace de Banach E, *les assertions suivantes sont équivalentes :*

a) E *ne contient aucun sous-espace isomorphe à* $\ell^1(N)$.

b) Tout élément $x'' \in E''$ *est Lusin-mesurable sur la boule unité* K *de* E', *munie de la topologie faible* $\sigma(E',E)$.

4. APPLICATIONS A L'ETUDE DES FONCTIONS VECTORIELLES SCALAIREMENT MESURABLES A VALEURS DANS UN ESPACE $C_S(K)$.

Pour étudier de telles fonctions, nous utilisons en suivant une idée de (7), les propriétés de compacité des ensembles de fonctions numériques mesurables. Du théorème (2.4) on déduit déjà, de manière classique, le résultat suivant (valable bien entendu moyennant l'hypothèse du continu et la condition de cardinalité sur la tribu Σ).

(4.1) THEOREME.

a) Tout ensemble A *de fonctions mesurables, compact dans* R^Ω *est précompact pour la topologie* $\mathcal{T}_m$ *de la convergence en mesure.*

b) Si de plus A *est séparé pour la topologie* $\mathcal{T}_m$, *alors* A *est compact pour* $\mathcal{T}_m$ *et sur* A *ces deux topologies compactes coïncident.*

Notons que M. TALAGRAND a montré dans (11) qu'il est en fait inutile de supposer que card $\overline{\Sigma} \leqslant 2^{\aleph_0}$.

Considérons maintenant une application $\vec{f} : (\Omega,\Sigma,\mu) \to C_S(K)$ et pour tout point x de K désignons par f_x l'application $\Omega \to R$ définie par $f_x(\omega) = \vec{f}(\omega)(x)$. On notera enfin par $C_U(K)$ l'espace C(K) muni de la topologie de la convergence uniforme sur K. Signalons encore que tous les résultats qui suivent sont valables en admettant l'hypothèse du continu (ou l'axiome, plus faible, de Martin) ; toutefois le théorème (4.1) de compacité étant valable, sans hypothèse particulière, pour des ensembles convexes (voir par exemple, TORTRAT (10)), il en est de même pour le théorème (4.8).

(4.2) THEOREME.

Soit $\vec{f} : (\Omega,\Sigma,\mu) \to C_S(K)$ *une fonction scalairement mesurable. On suppose que l'ensemble* $H = \{f_x\}$, $x \in K$, *est séparé pour la topologie de la convergence en mesure. Alors la fonction* $\vec{f}$ *est Bochner-mesurable, à valeurs dans l'espace* $C_U(K)$.

Ce théorème permet à son tour d'étudier les mesures de Baire, ou de Borel, sur l'espace $C_S(K)$. Désignons par $\mathfrak{B}a(\mu)$ (resp. $\mathfrak{B}(\mu)$) la tribu de Baire (resp. de Borel) complétée pour la mesure μ et introduisons la définition suivante :

(4.3) DEFINITION.

Soit μ *une mesure de Baire (resp. de Borel) bornée sur l'espace* $C_S(K)$ *et soit* F *un élément de* $\mathfrak{B}a(\mu)$ *(resp.* $\mathfrak{B}(\mu)$*). On dit que* F *est un porteur de* μ *si l'on a* $\mu(F) = 1$ *et si pour tout conoyau* $Z \subset C_S(K)$ *tel que* $F \cap Z \neq \emptyset$, *on a* $\mu(Z \cap F) > 0$.

On peut maintenant établir :

(4.4) THEOREME.

Soit μ *une mesure de Baire (resp. de Borel) bornée sur l'espace* $C_S(K)$ *et admettant un porteur* $F \in \mathfrak{B}a(\mu)$ *(resp.* $\mathfrak{B}(\mu)$*). Alors* μ *est de Radon sur l'espace* $C_U(K)$.

PREUVE.

Il est clair que l'application identique $\vec{I} : (F, \mathfrak{B}a(\mu),\mu) \to F \subset C_S(K)$ satisfait aux conditions du théorème (4.2). L'ensemble $\vec{I}(F)$ est alors séparable pour la norme de $C_U(K)$ et la mesure μ est donc de Radon puisque $\vec{I}(F)$ est polonais.

Soit T un espace complètement régulier ; rappelons que l'on désigne respectivement par $M_\sigma(T)$, $M_\tau(T)$ et $M_t(T)$ l'ensemble des mesures de Baire bornées sur T, des mesures de Borel bornées et τ-régulières, des mesures de Radon bornées sur T. Le théorème

(4.4) permet ainsi de retrouver, dans le cas particulier d'un espace $C_S(K)$ le résultat suivant établi par L. SCHWARTZ (10) pour un Banach affaibli.

(4.5) THEOREME.

Toute mesure bornée μ *qui est* τ*-régulière sur l'espace* $C_S(K)$ *est de Radon sur* $C_U(K)$. *En d'autres termes on a :*

$$M_\tau(C_S(K)) = M_t(C_S(K)) = M_t(C_U(K)).$$

En utilisant le théorème (4.1) et un théorème de relèvement dû à A. IONESCU-TULCEA (6) on peut également établir :

(4.6) THEOREME.

Soit L *un compact de l'espace* $C_S(K)$ *et soit* $\vec{f} : (\Omega,\Sigma,\mu) \to L$ *une fonction scalairement mesurable. Il existe alors une fonction* $f_\circ : \Omega \to L$ *qui est Bochner-mesurable et faiblement équivalente à* $\vec{f}$ *(i.e. pour tout* $x \in K$ *, on a* $f_x = f_{\circ x}$ μ*-presque partout).*

On en tire encore le résultat suivant pour les mesures de Baire sur L :

(4.7) THEOREME.

Soit L *un compact de l'espace* $C_S(K)$. *Alors toute mesure de Baire bornée sur* L *est de Radon. En d'autres termes, on a* $M_\sigma(L) = M_t(L)$.

On a évidemment le même résultat pour les compacts d'un espace de Banach affaibli. Ainsi, on peut montrer que le théorème établi dans (9) est en fait valable sans hypothèse de cardinalité sur l'espace métrisable T. De manière plus précise on a :

(4.8) THEOREME.

Soit T *un espace topologique métrisable ; les assertions suivantes sont équivalentes :*

a) T *est radonien ;*

b) T *est Radon-universellement mesurable.*

PREUVE.

Il suffit de montrer b) $\Longrightarrow$ a) ; la preuve résulte du théorème (4.7) et du fait que tout espace topologique métrisable T se plonge dans une partie faiblement compacte d'un espace $\ell^2(I)$ (résultat dû à W. SCHACHERMAYER (9)).

Comme autre application du théorème (4.1), on peut encore citer le résultat suivant :

(4.9) THEOREME.

Soit L *un ensemble compact de l'espace* $C_S(K)$ *qui est le porteur d'une mesure de Baire* μ *(au sens de la définition (4.3)) ; alors* L *est métrisable pour la topologie de* $C_S(K)$.

PREUVE.

L'application $I : K \to R^L$ définie par $I(x) = (f(x))$, $f \in L$, étant continue, l'ensemble I(K) est compact et de plus, il est clair que I(K) est séparé pour la topologie de la convergence en mesure. Ainsi I(K) est métrisable et il existe une suite $(x_n) \subset K$ telle que la suite $(I(x_n))$ soit dense dans I(K). On vérifie ensuite que la suite (x_n) sépare les points de L et il en résulte que L est métrisable pour la topologie de la convergence simple sur K.

BIBLIOGRAPHIE.

(1) D. BUCCHIONI et A. GOLDMAN, *Sur la convergence presque partout des suites de fonctions mesurables*, C.R. Acad. Sc. Paris, 283, 1976, p. 1087-1089.

(2) D. BUCCHIONI et A. GOLDMAN, *Sur la convergence presque partout des suites de fonctions mesurables*, Canad. J. of Math., à paraître.

(3) D.H. FREMLIN, *Pointwise compact subsets of measurable functions*, Manuscripta Math., 15, 1975, p. 219-242.

(4) A. GOLDMAN, Thèse d'Etat (en préparation).

(5) R. HAYDON, *Some more characterizations of Banach spaces containing* ℓ^1, Math. Proc. Camb. Phil. Soc., 1976, 80, p. 269-276.

(6) A. IONESCU-TULCEA and C. IONESCU-TULCEA, *Topics in the theory of liftings*, Berlin-Heidelberg-New-York, Springer, 1969.

(7) A. IONESCU-TULCEA, *On pointwise convergence, compactness and equicontinuity in the lifting topology I*, Z. Wahrscheinlichkeitstheorie verw. Geb., 26, 1973, p. 197-205.

(8) H.P. ROSENTHAL, *Pointwise compact subsets of the first Baire-class, with some applications to the Banach theory*, Aarhus Universitet, Mathematik Institut, Various Publications, series n° 24, 1975, p. 176-187.

(9) W. SCHACHERMAYER, *Eberlein-compacts et espaces de Radon*, C.R. Acad. Sc. Paris, 284, 1977, p. 405-407.

(10) L. SCHWARTZ, *Certaines propriétés des mesures sur les espaces de Banach*, Séminaire Ecole Polytechnique, exposé n° XXIII, année 1975-1976.

(11) M. TALAGRAND, *Solution d'un problème de A. IONESCU-TULCEA*, C.R. Acad. Sc. Paris, 283, 1976, p. 975-978.

ON THE COMPLETION OF VECTOR MEASURES

by

Davor Butković (Zagreb)

The purpose of this note is to relate various integration theories of vector-valued measures on locally compact spaces. We construct a null-completion of countably additive set functions which has properties analogous to the properties of the Lebesgue completion in the scalar case. Completing a regular Borel measure in such a way we obtain a space of integrable functions which coincides with the space of functions integrable with respect to the corresponding Radon vector measure. The integrals involved are those introduced by D. R. Lewis [7] and by E. Thomas [9].

Notations. Throughout, R will be a δ-ring of subsets of a set T and X will be a Hausdorff locally convex space. We denote by X', X^*, $\hat{X}$ the topological dual, the algebraic dual and the completion of X, respectively. For any continuous seminorm p on X we write X_p for $(X/p^{-1}(0))\hat{}$ and we denote the canonical projection $X \to X_p$ by π_p. By R^{loc} we denote the σ-algebra of sets $A \subset T$ satisfying $A \cap B \in R$ for every $B \in R$. Given $B \in R$ we denote by R_B the σ-algebra of sets A satisfying $A \in R$ and $A \subset B$. Let $m:R \to X$ be a set function. Given a continuous seminorm p on X we denote $\pi_p \circ m$ by m_p. If $A \in R$ we write m_A instead of $m|R_A$.

1. Null-completions

1.1. Let $m:R \to X$ be a measure i.e. a countably additive set function. As usual, call a set $A \in R$ m-null if $m(B) = 0$ for every $B \subset A$, $B \in R$. Denote by $N(m)$ the class of all m-null sets of R. Further, denote by $\overline{R}^m$ the δ-ring of all sets of the form $C = A \cup B$ where $A \in R$ and B is a subset of some m-null set; the measure $\overline{m}$ on $\overline{R}^m$ defined by $\overline{m}(C) = m(A)$ is called the Lebesgue completion of m.

1.2. One has $x' \circ \overline{m} = \overline{x' \circ m} \mid \overline{R}^m$ $(x' \in X')$. On the other hand, for every $x' \in X'$, $\overline{x' \circ m}$ is well defined on the δ-ring $\bigcap_{x' \in X'} \overline{R}^{x' \circ m}$. Denote this ring by $\tilde{R}^m$. If X is a normed space, $\tilde{R}^m = \overline{R}^m$, namely, in this case given $A \in R$ there exists the "Rybakov functional" $x_A' \in X'$ such that $N(m_A) = N(x_A' \circ m_A)$ [8, p.250] and consequently, $(R_A)^{-m_A} = (R_A)^{\sim m_A} = (R_A)^{-x_A' \circ m_A}$ for every $A \in R$.

1.3. In general case it can happen that $\tilde{R}^m$ is larger than $\overline{R}^m$. To see this take $T = [0,1]$. Let R be the algebra of Borel subsets of $[0,1]$, and let $X = \mathbb{R}^T$ with the product topology. Define $m:R \to X$ by $m(A) = \chi_A$ (where χ_A is the characteristic function of $A \in R$). Then $\overline{R}^m = R$. However, it is easy to see that $\tilde{R}^m$ coincides with the power set of T.

1.4. Therefore it is natural to look for an extension of m on $\tilde{R}^m$ with properties analogous to the properties of $\overline{m}$. Call a set function $\tilde{m}:\tilde{R}^m \to \hat{X}$ the scalar null-completion of m if $x' \circ \tilde{m} = \overline{x' \circ m} | \tilde{R}^m$ for every $x' \in X'$. The scalar null-completion is again a measure which extends m (and even extends $\overline{m}$). One has the following existence and uniqueness

1.5. THEOREM. Every measure $m:R \to X$ has a unique scalar null-completion.

1.6. Proof. Let $A \in \tilde{R}^m$. One can find a $D \in R$ such that $A \subset D$. Further, for every $x' \in X'$ we have $A = B_{x'} \cup C_{x'}$, where $B_{x'} \in R$ and $C_{x'} \in N(\overline{x' \circ m_D})$. Hence, for every finite set $H \subset X'$, we have $A = B \cup C$ with $B = \bigcup_{x' \in H} B_{x'} \in R_D$ and $C = \bigcap_{x' \in H} C_{x'} \in N(\overline{x' \circ m_D})$ for every $x' \in H$. It follows that $\bigcap_{x' \in H} x'^{-1}((\overline{x' \circ m_D})(A)) \neq \emptyset$ for any finite $H \in X'$. Since m_D has a conditionally weakly compact range in $\hat{X}$ and since for any $x' \in X'$ the set $x'^{-1}((\overline{x' \circ m_D})(A))$ is closed in the weak topology on $\hat{X}$, we have $\bigcap_{x' \in H} x'^{-1}((\overline{x' \circ m_D})(A)) \neq \emptyset$. It is easy to see that this intersection contains exactly one element, say x_A. Define $\tilde{m}(A) = x_A$. By construction, one has $(x' \circ \tilde{m})(A) = (\overline{x' \circ m})(A)$ for every $x' \in X'$.

1.7. Since X_p are Banach spaces, $\tilde{R}^{m_p} = \overline{R}^{m_p}$ and $(m_p)^{\sim} = (m_p)^{-}$ for every continuous seminorm p on X. A simple computation gives $\tilde{R}^m = \bigcap_{p \in P} \overline{R}^{m_p}$, $N(\tilde{m}) = \bigcap_{p \in P} N((m_p)^{-})$, and $(\tilde{m})_p = (m_p)^{-} | \tilde{R}^m$ for every family P of seminorms p which determines the topology of X. It follows that

$$\tilde{m}(A) = \bigcap_{x' \in X'} x'^{-1}((\overline{x' \circ m})(A)) = \bigcap_{p \in P} \pi_p^{-1}((\overline{\pi_p \circ m})(A))$$

for every $A \in \tilde{R}^m$.

1.8. Now we turn to the integration theory. Let f be a scalar function on T. Given a scalar measure λ on R, by the integral of $f|A$ on $A \in R$ with respect to λ_A we always mean the integral in the sense of [5]. f is called <u>λ-integrable</u> on T, $f \in L^1(\lambda)$, if $\sup_{A \in R} \int |(f|A)| \, d|\lambda_A| < +\infty$; $\int f \, d\lambda$ will be a "essential" integral in the sense of [1]. Given a vector measure $m:R \to X$, f is called <u>m-weakly integrable</u> if $f \in L^1(x' \circ m)$ for every $x' \in X'$. In this case the <u>weak integral</u> $(w) \int f \, dm \in X'^*$ is defined by

$< (w)\int f\,dm, x' > = \int f\,d(x'\circ m)$ $(x' \in X')$. If $A \in R^{loc}$ we write $(w)\int_A f\,dm$ for $(w)\int f\,\chi_A\,dm$. In all these cases f is also <u>R^{loc}-measurable</u> in the sense of [7].

1.9. PROPOSITION. Let $m:R \to X$ be a measure and let f be a scalar function on T. Then

(i) f is $(\tilde{R}^m)^{loc}$-measurable iff it is $(\overline{R}^{x'\circ m})^{loc}$-measurable for every $x' \in X'$,

(ii) f is $\tilde{m}$-weakly integrable iff it is $\overline{x'\circ m}$-integrable for every $x' \in X'$; in this case we have

$$\int_A f\,d(x'\circ \tilde{m}) = \int_A f\,d(\overline{x'\circ m})$$

for every $A \in (\tilde{R}^m)^{loc}$ and every $x' \in X'$.

1.10. Proof. The statement (i) is obvious by the definitions of measurability and null-completions. To prove (ii) observe that for any δ-ring R_1 and for any scalar measures λ, λ_1 on R, R_1 respectively, satisfying $R \subset R_1 \subset \overline{R}^\lambda$ and $\lambda_1 = \overline{\lambda}|R_1$, we have $f \in L^1(\lambda_1)$ iff f is $(R_1)^{loc}$-measurable and belongs to $L^1(\overline{\lambda}_1) = L^1(\overline{\lambda})$. Applying this to $R \subset \tilde{R}^m \subset \overline{R}^{x'\circ m}$ and to $x'\circ \tilde{m} = \overline{x'\circ m}\,\big|\,\tilde{R}^m$ where $x' \in X'$ it follows that $f \in L^1(\overline{x'\circ m})$ for every $x' \in X'$ iff $f \in L^1(x'\circ \tilde{m})$ for every $x' \in X'$.

1.11. Our scalar null-completion behaves well in the integration theory developed by D. R. Lewis [6], [7]. This has been shown in [3] where we allow the integrals to take values in $\hat{X}$ and we define the function f to be <u>m-integrable</u>, $f \in L^1(m)$, if it is m-weakly integrable and $(w)\int_A f\,dm \in \hat{X}$ for all $A \in R^{loc}$. For example, one can prove (see [3]) that the following are equivalent:

(i) $f \in L^1(\tilde{m})$,

(ii) f is $\tilde{m}$-weakly integrable and $(w)\int_A f\,d\tilde{m} \in \hat{X}$ for every $A \in R^{loc}$, and

(iii) $f \in L^1((m_p)^-)$ for each continuous seminorm p on X.

2. Borel and Radon measures

2.1. From now on T will be a locally compact Hausdorff space. We denote by B(T) the δ-ring generated by all compact subsets of T. A measure defined on B(T) is called a Borel measure (on T). Borel measures are uniquely determined by their values on compact sets [4, III.14.4, Prop.17, p.297]. A Borel measure $m:B(T) \to X$ is called regular if, for every continuous seminorm p on X, m_p is regular in the sense of [4, III.15.3, p.304] (similarly for inner and outer regularity). We simplify notations by writing $\tilde{B}^m(T)$ for $B(T)^{\sim m}$, $\tilde{B}^m(T)^{loc}$ for $(B(T)^{\sim m})^{loc}$,etc.

2.2. Denote by K(T) the space of continuous scalar functions with compact supports on T, endowed by the inductive limit topology [1, III.1.1, p.40]. A Radon measure on T is a continuous mapping $\mu:K(T) \to X$; integrability with respect to a scalar Radon measure is the "essential" integrability of [1, V]. A vector-valued Radon measure μ is called extendible [9, 3.1, p.100] if every bounded scalar Borel function f with a compact support is μ-integrable in the sense of [9, 1.27, p.77] ($f \in L^1(\mu)$). Extendible measures μ are uniquely determined by their values $\mu(K) = \int \chi_K \, d\mu$ for compact $K \subset T$ (use [1, IV.4.10, Cor.3 of Prop.19, p.163] for $x' \circ \mu$, $x' \in X'$).

2.3. An extendible Radon measure μ and a Borel measure m on T with values in a complete space are corresponding if $\mu(K) = m(K)$ for every compact $K \subset T$ (analogously for scalar Radon and Borel measures). In the scalar case correspondence gives a bijection between the set of all Radon measures and the class of regular Borel measures [1, IV.4.11, Th.5, p.164]. By the regularity of m and by the Urysohn lemma one has $m(A) = \int \chi_A \, d\mu$ for $A \in B(T)$ and $\mu(f) = \int f \, dm$ for $f \in K(T)$. The equality of these integrals with respect to corresponding scalar measures μ and m for B(T)-simple functions and for functions from K(T) implies that the measures $|\mu|$ [1, III.1.6, (12), p.55] and $|m|$ (on B(T)) [7, 2.1, p.295] are also corresponding. In the vector case we have the following

2.4. PROPOSITION. Let T be locally compact and let X be complete. Then there exists a bijection between the set of X-valued extendible Radon measures and the set of X-valued regular Borel measures on T such that the measures from the established pairs are corresponding.

2.5. Proof. Given an extendible μ, $\chi_A \in L^1(\mu)$ for every $A \in B(T)$. Further, $m: A \mapsto \int \chi_A \, d\mu$ is a set function from B(T) to X which is countably additive by the Orlicz-Pettis theorem. The regularity of $x' \circ m$, $x' \in X'$, gives the regularity of m_A, $A \in B(T)$ [6, 1.6, p.159]. Therefore m is innerly regular and, by [4, III.15.3, Prop.4, p.306], it is regular.

Conversely, let m be a regular Borel measure, and let $f \in K(T)$. Then $f \in L^1(m)$ by [6, 2.4, p.162]. Set $\mu(f) = \int f \, dm$. By [6,(3), p.160] μ is a Radon measure and, since $(x' \circ \mu)(f) = \int f \, d(x' \circ m)$, the scalar measures $x' \circ \mu$ and $x' \circ m$ are corresponding for every $x' \in X'$. Therefore $\int \chi_K \, d(x' \circ \mu) = \langle m(K), x' \rangle$ for every compact $K \subset T$ and every $x' \in X'$. By [9, 3.3(b), p.101], μ is extendible, $\chi_K \in L^1(\mu)$, and $\int \chi_K \, d\mu =$ $= (w) \int_K d\mu = m(K)$.

2.6. Let the space X be normed, let $\mu: K(T) \to X$ be a Radon measure, and let f be a scalar function on T. Then μ-measurability is defined via the semivariation $\mu^\bullet$ [9, 1.1, p.65] by the Luzin property: a function f is called <u>μ-measurable</u> if for every compact $K \subset T$ and every $\varepsilon > 0$ there exists a compact $K_1 \subset K$ such that $\mu^\bullet(K-K_1) < \varepsilon$ and such that $f|K_1$ is continuous. If X is locally convex a function f is called μ-measurable if, for every continuous seminorm p on X, f is μ_p-measurable where $\mu_p = \pi_p \circ m$ [9, 1.27, p.77], in the case of extendible μ, this is equivalent to the $x'\circ\mu$-measurability of f for every $x' \in X'$ [9, 3.5, p.102].

2.7. Let μ, m be two corresponding vector-valued measures (with $m: B(T) \to X$). Then $(\mu_p)^\bullet$ and $\|m_p\|$ [7, p.296] coincide on B(T) since, by [9, 1.26, p.76], $|x_p'\circ\mu_p|(A) = |x_p'\circ m_p|(A)$ for every $x_p' \in (X_p)'$. It follows that the space of μ-measurable functions coincides with the space of functions which for every continuous seminorm p on X have the Luzin property in terms of $\|m_p\|$. On the other hand, by our theorem 2.8 bellow, $\tilde{B}^m(T)^{loc}$ can be characterized in terms of μ-measurability (cf. [10, 2.3, p.21]).

2.8. THEOREM. Suppose that an extendible Radon measure μ on T and a Borel measure m are corresponding. Then

(i) the class of μ-measurable scalar functions coincides with the class of $\tilde{B}^m(T)^{loc}$-measurable scalar functions,

(ii) $L^1(\mu) = L^1(\tilde{m})$, and

$$\int f\, d\mu = \int f\, d\tilde{m} \qquad (f \in L^1(\mu)).$$

2.9. Proof. First, consider the scalar case with $\bar{m}$ instead of $\tilde{m}$. By [1, IV.4.6, Th.4, p.152] and by the regularity of m the class of conditionally compact μ-integrable sets coincides with $\bar{B}^m(T)$ and $\bar{m}(A) = \int \chi_A\, d\mu$ for $A \in \bar{B}^m(T)$. By [1, IV.5.5, Prop.8, p.180], (i) follows. Each bounded function with compact support is μ-, resp. $\bar{m}$-integrable iff it is μ-, resp. $\bar{B}^m(T)^{loc}$-measurable, hence, by monotone convergence theorems we obtain $L^1(\mu) = L^1(\bar{m})$. The last statement follows by the equality of integrals on $\bar{B}^m(T)$-simple functions.

Now, consider the vector case. If $x' \in X'$, $x'\circ\mu$-measurability coincides with $\bar{B}^{x'\circ m}(T)^{loc}$-measurability, therefore, (i) follows by 1.9(i). By 1.9(ii) and by the Theorem in the scalar case, the class of μ-weakly integrable functions coincides with the class of $\tilde{m}$-weakly integrable functions, for such functions f we have $(w)\int_A f\, d\mu = (w)\int_A f\, d\tilde{m}$ for every $A \in \tilde{B}^m(T)^{loc}$.

Finally, $f \in L^1(\mu)$ iff f is μ-weakly integrable and $(w)\int_A f\, d\mu \in \hat{X}$ for every open $A \subset T$ [9, 3.11, p.106]. By [9, 1.22, p.74], $f \in L^1(\mu)$ iff $(w)\int_A f\, d\mu \in \hat{X}$ for every $A \in \tilde{B}^m(T)^{loc}$, and this is equivalent to the fact that $f \in L^1(\tilde{m})$.

2.10. One can generalize the above theorem to measures on arbitrary Hausdorf topological spaces [3]; in this case one has to replace the extendible measures with vector-valued premeasures [2].

REFERENCES

1. N. BOURBAKI, Intégration, Chap. I-IV 2ème éd. 1965, Chap V 2ème éd. 1967, Hermann, Paris.
2. D. BUTKOVIĆ, Integration with respect to Radon vector premeasures, Glasnik matematički, vol. 11 (31) (1976), 263-289.
3. -, On Borel and Radon vector measures, Glasnik matematički, to appear.
4. N. DINCULEANU, Vector measures, VEB Deutscher Verlag der Wissenschaften, Berlin, 1966.
5. P. R. HALMOS, Measure theory, Van Norstand, New York, 1950.
6. D. R. LEWIS, Integration with respect to vector measures, Pacific J. Math., vol. 33 (1970), 157-165.
7. -, On integrability and summability in vector spaces, Illinois J. Math., vol 16 (1972), 294-307.
8. V. I. RYBAKOV, K teoreme Bartla-Danforda-Švarca o vektornyh merah, Matematičeskie zametki, vol. 7, 2 (1970), 247-254 *(in russian)*.
9. E. THOMAS, L'intégration par rapport à une mesure de Radon vectorielle, Ann. Inst. Fourier, Grenoble, vol. 20, 2 (1970), 55-191.
10. -, On Radon maps with values in arbitrary topological vector spaces, and their integral extensions, Yale University, 1972.

University of Zagreb, Yugoslavia
(Elektrotehnički fakultet, Unska b.b.)

STOCHASTIC PROCESSES AND COMMUTATION RELATIONSHIPS

S. D. Chatterji

§1. INTRODUCTION

The purpose of the present paper is to offer an introductory sketch of the analysis of second-order stationary stochastic processes by means of standard Hilbert space theory and a fundamental theorem due to Mackey [13(a)] (in the case of the group $\mathbb{R}^n$ due to Stone (irreducible case) and von Neumann). The use of the latter theorem in this context is recent (cf. Tjøstheim [17], Gustafson and Misra [4]) although the basic imprimitivity relation (6) had been noticed, in this situation, by other authors (cf. Hanner [5], Kallianpur and Mandrekar [9]). The imprimitivity relation (6) is a far-reaching generalization of commutation relationships of quantum mechanics (cf. [18] and is the source of numerous other developments such as the induced representations (cf.[13(b)]), the Beurling theory of invariant subspaces (cf.[6]) and scattering theory (cf.[10]). We hope to include these and other aspects in a comprehensive survey in a later publication [1].

In §2 we sketch the general set-up; an even more general situation, currently under much investigation (cf.[2], [14(c)]) is indicated in the last section. This would permit us to cover the case of the generalized stationary fields. The stationary fields correspond to situations where one has a unitary group as "propagator". Interesting non-stationery fields correspond to non-unitary propagators and are related to Sz.-Nagy dilation theory (cf.[14(c)]). Our interest in pointing out the present approach is that this permits us, in principle, to analyse regularity of processes on more general groups. It is our hope that the results of Muhly [15] on the Bohr group would be deducible from our methods and that far-reaching generalisations of the theory, so far strictly limited to $\mathbb{R}$ or $\mathbb{Z}$, would be possible for very general groups where a natural definition of regularity would seem to be the one indicated in §4.

In §3, we indicate how our method applies to the classical case of a discrete parameter stationary process. The analysis is based on ideas of [4] and [17] in the case of the group $\mathbb{R}$. The group $\mathbb{Z}$ is perhaps more instructive in this analysis in so far as the dual of $\mathbb{Z}$ is a group different from $\mathbb{Z}$ itself.

There are no new or difficult results in this paper. It is more of a manifesto for and a promise of future developments.

§2. GENERALITIES

A general formulation of the notion of a stationary process can be given as follows. Let S be a set, G a semi-group acting on S and ξ a map from S to a Hilbert space (real or complex) H. Let the scalar product of two elements x and y of H be denoted by $(x|y)$. Then ξ is said to define a stationary process if

$$(\xi(s)|\xi(t)) = (\xi(gs)|\xi(gt) \tag{1}$$

for all s,t in S and $g \in G$. In many practical problems, S is a nice manifold, G is a group of automorphisms of S and H is the space of square-integrable random variables on some probability space. The case $S=G=\mathbb{R}$ (or Z) where G acts on itself by translation is of particular importance. The function $(s,t)\to(\xi(s)|\xi(t)) = C(s,t)$ is called the covariance function of the process. The closed linear manifold (or subspace) $H(\xi)$ generated by $\{\xi(s)|s \in S\}$ (in symbols: $H(\xi) = \text{clm}\ \{\xi(s)|s \in S\}$) is called the time domain of the process in case $G=\mathbb{R}$ or $\mathbb{R}_+$ (or Z or $\mathbb{N}$) is thought of as time; in the general case, $H(\xi)$ should probably be called the process space. Without loss of generality, we can and shall assume that $H(\xi) = H$. The purpose of the time domain (or process space) analysis is to obtain a reasonable representation of $H(\xi)$ as a set of functions on S. The spectral domain (or space) in this generality, can be considered as a set of functions on a suitable dual space $\tilde{G}$ of G (typically a space of characters or representations of G) obtained by an analysis (harmonic analysis) of the covariance function C which is a non-negative definite kernel on SxS. The purpose of the spectral domain analysis is to obtain an isomorphism between the process space and the spectral space and then to use the powerful tools of harmonic analysis on the spectral space.

If G is a group acting transitively on S then S is equivalent to the left coset space G/K where K is a suitable subgroup of G (e.g. K can be taken as the set of $k \in G$ such that $ks_o = s_o$ for some fixed $s_o \in S$); hence C can be considered to be a function on (G/K)x(G/K) invariant under the action of G. If further K is a normal subgroup of G, then $C(g_1K, g_2K) = \rho(g_2^{-1}g_1 K)$ where ρ is a non-negative definite function on the group G/K; this is verified immediately from (1). In particular, if S=G then K can be taken to be the trivial proper subgroup of G and $C(g_1,g_2) = \rho(g_2^{-1}g_1)$ for some non-negative definite function ρ on G—— also called the <u>covariance function</u>.

Suppose now that $H(\xi)=H$, S=G, a group and define

$$U_g\xi(s) = \xi(gs) \tag{2}$$

An easy calculation using (1) yields that

$$\left\| \sum_{j=1}^{n} a_j \, \xi(s_j) \right\|^2 = \sum_{1 \leq j,k \leq n} a_j \, \bar{a}_k \, (\xi(gs_j) | \xi(gs_k))$$

for any complex a_j and elements g, s_j from G. This shows that U_g can be extended by linearity and continuity to be an unitary operator on H; further, from (2), $U_{g_1} U_{g_2} = U_{g_1 g_2}$. In other words, $g \to U_g$ is an unitary representation of the group G corresponding to the non-negative definite function ρ (cf. the Gelfand-Raikov theorem). Further, the representation U possesses a cyclic vector $\xi_o = \xi(e)$ where e is the unit element of G since $\xi(g) = U_g \xi_o$, $g \in G$, generate H. Also

$$\rho(g) = C(g,e) = (U_g \xi_o | \xi_o) \tag{3}$$

It is also clear in this case (S = G,a group) that two stationary processes ξ and η are unitarily equivalent in an obvious sense iff they possess the same covariance function ρ or iff the relevant unitary groups and cyclic vectors correspond under an unitary map. Thus, from a group theoretical point of view, a stationary process can be seen as an unitary representation with a prescribed cyclic vector; from an analytic view point, it is summed up by the function ρ.

Let us now suppose further that S = G is a locally compact abelian group. If ξ is a continuous map (which is the case iff ρ is continuous or if H is separable and ρ is measurable i.e. $\xi: G \to H$ is weakly measurable) then there exists a projection-valued measure E on the Borel sets of $\hat{G}$, the dual group of G, such that

$$\left.\begin{aligned} U_g &= \int_{\hat{G}} \lambda(g)\, E(d\lambda) \\ \xi(g) = U_g\, \xi_0 &= \int_{\hat{G}} \lambda(g)\, E(d\lambda)\, \xi_0 \\ \rho(g) = (U_g \xi_0 \,|\, \xi_0) &= \int_{\hat{G}} \lambda(g)\, (E(d\lambda)\, \xi_0 \,|\, \xi_0) \end{aligned}\right\} \quad (4)$$

The H-valued measure $B \to E(B)\xi_0$ on the Borel sets of $\hat{G}$ is of a well-known type (called c.a.o.s = countably additive, orthogonally scattered by Masani [14(b)]). By the elementary theory of such measures and (4), the space $L^2(\hat{G},\mu)$, where μ is the positive measure $B \to (E(B)\xi_0 \,|\, \xi_0) = \mu(B)$ (called the spectral measure of the process), is isomorphic to $H = H(\xi)$ under the unitary map which sends the function $\lambda \to \lambda(g)$ on $\hat{G}$ to $\xi(g) \in H$. This is the standard analytic tool of mapping the process space to the spectral space.

§3. CASE S=G=Z .

Here $\hat{G}$ can be identified with the unit circle T of the complex plane (which we shall confound with the interval $[0,2\pi[$). The analysis of the previous section can be sharpened here in a remarkable way if we suppose that the process ξ is purely non-deterministic or regular i.e.

$$\bigcap_n H_n = \{0\} \quad (5)$$

where $H_n = \text{clm}\,\{\xi(g) \,|\, g \leq n\}$. The linear ordering of Z , the monotonicity of the map $n \to H_n$ and (5) immediately yields a projection-valued measure on all subsets of Z as follows : let P_n be the orthogonal projection on the closed linear manifold $H_n \ominus H_{n-1}$ (also called the nth innovation subspace). Define, for $A \subset Z$,

$$P(A) = \sum_{n \in A} P_n$$

It can be easily seen that P is a projection-valued measure (i.e.

$A \to (P(A)x|y)$ is countably additive for all x,y in H) and that $P(Z) = I$. The projection-valued measure P intertwines with the unitary representation $g \to U_g$ of the previous section according to the following formula, known as the <u>imprimitivity relation</u> :

$$U_g \; P(A) \; U_{-g} = P(A+g) \tag{6}$$

for all $A \subset Z$ and $g \in Z$. The verification of (6) i this case is done most easily by taking one point sets A and elements $\xi(m) \in H$. We shall now prove that $\{U_g, g \in Z \; ; \; P(A), A \subset Z\}$ forms an irreducible family of operators. The shortest proof of this is that of Tjøstheim [17(a)] and can be based on the Mackey-von Neumann-Stone theorem referred to before. According to this, in the general form given in Mackey [13(a)], (cf. also [12], [7]) any unitary representation U of a (separable) locally compact group G and a projection valued measure P on the Borel subsets of G which satisfy (6) is unitarily equivalent to the direct sum of a certain number (at most denumerable) of canonical objects on $L^2(G)$ viz. the (left) regular representation and the canonical spectral measure of multiplication by characteristic functions. Since these latter have multiplicity one (cf. [7], p.316) and the U in question here has a cyclic vector (i.e. is simple), irreducibility follows. By virtue of the same theorem, we conclude now that there exists an unitary map $V: H \to L^2(Z)$ such that $(g, x \in Z, A \subset Z)$

$$\left.\begin{aligned} (VU_gV^{-1})f(x) &= f(x-g) \\ (VP(A)V^{-1})f(x) &= \varphi_A(x)f(x) \end{aligned}\right\} \tag{7}$$

where φ_A is the characteristic function of the set A. Let $V\xi_0 = f_0 \in L^2(Z)$ (the space of square-summable sequences on Z) ; if A is the set of integers less than or equal to zero, we have,

$$\begin{aligned} \varphi_A \cdot f_0 &= VP(A)V^{-1} f_0 \\ &= VP(A)\xi_0 \\ &= V\xi_0 \qquad\qquad (\text{since } \xi_0 \in H_0 \ominus H_{-1}) \\ &= f_0 \end{aligned}$$

so that $f_0(x) = 0$ if $x \geq 1$.

Thus our abstract situation is unitarily equivalent to the following :

$$(*)\qquad \begin{cases} H = L^2(Z)\,, \\ (\tilde{U}_g f)(x) = f(x-g)\,, \quad g \in Z\,, \\ \text{cyclic vector } f_0 \text{ such that } f_0(x) = 0 \text{ if } x \geqslant 1. \end{cases}$$

This clearly is a characterization of a regular process (in the case S=G=Z) : one which is unitarily equivalent to (*) for some f_0 of the type indicated.

All this analysis is in the process space. Passage to the spectral space allows the application of the profound machinery of harmonic analysis which we now proceed to illustrate.

Let

$$\tilde{U}_n = \tilde{U}^n = \int_0^{2\pi} \exp(in\lambda)\ \tilde{E}(d\lambda)$$

where $\tilde{E}$ is a projection-valued measure on the Borel sets of T (with projections in $L^2(Z)$). Clearly $\tilde{E}$ is unitarily equivalent to E of (4); also :

$$\begin{aligned} \rho(n) &= (U^n \xi_0 \mid \xi_0) \\ &= (\tilde{U}^n f_0 \mid f_0) \\ &= \int_0^{2\pi} \exp(in\lambda)\ \mu(d\lambda) \end{aligned}$$

$$\text{(where } \mu(B) = (\tilde{E}(B) f_0 \mid f_0)_{L^2(Z)} \geqslant 0 \ )$$

$$= \sum_x f_0(x-n)\ \overline{f_0(x)}$$

(where $\sum_x$ stands for $\sum_{-\infty}^{\infty}$) ;

$$\text{if } \psi_0(\lambda) = \sum_x f_0(x)\ \exp(i\lambda x) \tag{8}$$

$$= \sum_{x \leqslant 0} f_0(x)\ \exp(i\lambda x)\,,$$

then

$$\sum_x f_0(x-n) \exp(i\lambda x) = \psi_0(\lambda) \exp(in\lambda)$$

and the Parseval formula gives us that

$$\rho(n) = \int_0^{2\pi} \exp(in\lambda)\, \mu(d\lambda)$$

$$= \sum_x f_0(x-n) \overline{f_0(x)}$$

$$= \int_0^{2\pi} \exp(in\lambda) \, |\psi_0(\lambda)|^2 d\lambda$$

$$= \int_0^{2\pi} \exp(-in\lambda) \, |\psi_0(-\lambda)|^2 d\lambda \quad .$$

Thus $\mu(d\lambda) = |\psi_0(\lambda)|^2 d\lambda$ and μ, the spectral measure of the process, is equivalent to Lebesgue measure on T. In fact, since $\sum_x |f_0(x)|^2 < \infty$, and, from (8),

$$\psi_0(-\lambda) = \sum_{x=0}^{\infty} f_0(-x) \exp(i\lambda x),$$

it follows from classical theory that the function $\lambda \to \psi_0(-\lambda)$ is the boundary value of a function of the class H^2 in the unit disc (cf. [8],p.39) and is non-zero a.e. (Lebesgue measure). The same theory tells us ([8],p.53) that the function $\lambda \to |\psi(-\lambda)|^2 = p(\lambda)$ (also called the spectral density of the process) is such that

$$\int_0^{2\pi} \log\ p(\lambda)\ d\lambda \ > -\infty \quad .$$

This is the classical Kolmogorov-Wiener (Krein-Szegö) criterion for the regularity of a discrete parameter (weakly) stationary process (cf. [3],p.577).

The so-called Wold decomposition (cf. [3]) can be obtained directly from the process space analysis. Indeed, if $e_n \in L^2(Z)$ is the usual orthonormal basis (i.e. $e_n(x) = 0$ if $n \neq x$ and $e_n(x) = 1$ if $n = x, n \in Z$) then

$$f_o = \sum_{m \leq o} f_o(m)\, e_m \quad ,$$

$$\tilde{U}_n f_o = \sum_{m \leq o} f_o(m)\, e_{m+n}$$

$$= \sum_{k=-\infty}^{n} f_o(k-n)\, e_k \quad ,$$

$$\xi(n) = V^{-1} \tilde{U}_n f_o$$

$$= \sum_{k=-\infty}^{n} f_o(k-n)\, \delta_k \qquad (9)$$

where $\delta_k = V^{-1} e_k$ is an orthonormal basis for $H = H(\xi)$. The formula (9) is called a moving average representation or Wold-decomposition of ξ.

In case $\bigcap_n H_n = H_{-\infty} \neq \{0\}$, the process $\xi = \xi' + \xi''$ where ξ' is a regular process obtained by projecting ξ on the orthogonal complement of $H_{-\infty}$ and ξ'' is what is known as a purely deterministic process.The above analysis then applies to ξ'.

§4. FINAL REMARKS

The analysis of §3 carries over verbatim to the case $S = G = \mathbb{R}$ —— the so-called continuous parameter case. The multi-variate situation which consists essentially of replacing the one-dimensional subspace spanned by ξ_o by a general (closed) subspace D evolving under a unitary group is handled with equal ease (cf. [11],[17(a)]). Of course, in practice, D is often finite-dimensional and the multivariate process is given in terms of the evolution of a basis $(\xi^1(o),\ldots,\xi^q(o))$ of D under the unitary group U i.e. in terms of $(\xi^1(g),\ldots,\xi^q(g))_{g \in G}$ (cf.[14(a)]).

An interesting generalisation is that of considering ξ as a map from S into L(E,H), the space of continuous linear operators from a locally convex space E into a Hilbert space H. If E is a space of test functions on $\mathbb{R}^n$ then this theory includes the vector-valued generalized fields. (cf. [2]).

The notion of regularity seems somewhat difficult to generalize to groups which are not linearly ordered. One possibility is to demand that the process be unitarily equivalent to one generated by translations on the group. An analysis of this remains to be done. Other approaches are offered by the so-called interpolation theory (cf.[16],[19]). Some ad hoc possibilities are indicated in the case of $\mathbb{R}^n$ in [17(b)].

REFERENCES

[1]. Chatterji, S.D.
to be sumitted to Jahrbuch Überblicke Mathematik 1978, Bibliographisches Institut,Mannheim/Wien/Zürich.

[2]. Chobanyan, S.A, and Weron, A.
Banach space valued stationary processes and their linear prediction, Dissertationes Math., 125 (1975), 1-50.

[3]. Doob, J.L.
Stochastic Processes. Wiley (1953).

[4]. Gustafson, K. and Misra, B.
Canonical commutation relations of quantum mechanics and stochastic regularity, Letters in Mathematical Physics, 1 (1976), 275-280.

[5]. Hanner, O.
Deterministic and non-deterministic processes, Ark.Mat., 1 (1950), 161-177.

[6]. Helson, H.
Lectures on invariant subspaces. Academic Press (1964).

[7]. Hewitt, E. and Ross, K.A.
Abstract harmonic analysis II. Springer-Verlag (1970).

[8]. Hoffman, K.
Banach spaces of analytic functions. Prentice-Hall (1962).

[9]. Kallianpur, G. and Mandrekar, V.
Multiplicity and representation theory of purely

non-deterministic stochastic processes, Theor. Probability Appl., 10 (1965), 553-581.

[10]. Lax, P.D. and Phillips, R.S.
Scattering theory. Academic Press (1967).

[11]. Lewis, J.T. and Thomas, L.C.
A characterization of regular solutions of a linear stochastic differential equation, Z. Wahrscheinlichkeitstheorie verw. Gebiete, 30 (1974), 45-55.

[12]. Loomis, L.H.
Note on a theorem of Mackey, Duke Math.J.,19 (1952),641-645.

[13]. Mackey, G.W.
(a) A theorem of Stone and von Neumann, Duke Math.J., 16 (1949), 313-326.
(b) Induced representations of groups and quantum mechanics. Benjamin (1968).

[14]. Masani, P.
(a) Recent trends in multivariate prediction theory, Multivariate Analysis (P.R. Krishnaiah, Editor) Proc. Internat.Sympos. (Dayton, Ohio, 1965) Academic Press (1966), 351-382.
(b) Quasi-isometric measures and their applications, Bull. Amer. Math. Soc., 76 (1970), 427-528.
(c) Dilations as propagators of Hilbertian varieties (to be published).

[15]. Muhly, P.S.
The distant future, Indiana Univ. Math.J., 24 (1974) 149-159.

[16]. Salehi, H. and Scheidt, J.K.
Interpolation of q-variate stationary stochastic processes over a locally compact abelian group, J. Multivariate Anal., 2 (1972), 307-331.

[17]. Tjøstheim, D.
(a) A commutation relation for wide sense stationary processes, SIAM J. Appl. Math., 30 (1976), 115-122.
(b) Spectral representations and density operators for infinite-dimensional homogeneous random fields, Z. Wahrscheinlichkeitstheorie verw. Gebiete, 35 (1976), 323-336.

[18]. Varadarajan, V.S.
Geometry of quantum mechanics II, Van Nostrand (1970).

[19]. Weron, A.
(a) On characterizations of interpolable and minimal stationary processes, Studia Math., 49 (1974), 165-183.
(b) (with Makagon, A.) Wold-Cramèr concordance theorems for interpolation of q-variate stationary processes over locally compact abelian groups, J. Multivariate Anal., 6 (1976), 123-137.

Département de Mathématiques
Ecole Polytechnique Fédérale de Lausanne
61 Av. de Cour
1007 LAUSANNE
SWITZERLAND.

SOME RESULTS WITH RELATION TO THE CONTROL MEASURE PROBLEM

Jens Peter Reus Christensen

Department of Mathematics,
University of Copenhagen,
DK-2100 Copenhagen,
Denmark.

Abstract : ,, If a sequentially point continuous submeasure (Maharam submeasure) defined on a measurable space admits a control measure, then a Fubini condition is fulfilled for measurable subsets of the product of the space with the unit interval (with usual Lebesgue measure). The main result is that this Fubini condition is also sufficient for the existense of control measure. Some related results are proved.,,

The control measure problem is equivalent with a problem first stated by D. Maharam (see [3]). Her question inspired the authors joint work with Wojchiech Herer (see [2]) and the present paper is a continuation of this investigation.

Let $(X, \mathcal{B})$ be a measurable space. A set function

$$\varphi : \mathcal{B} \dashrightarrow [0,\infty[$$

is called a submeasure if it satisfies the 3 conditions

1) $A \subseteq B \Rightarrow \varphi(A) \leq \varphi(B)$.
2) $\varphi(\emptyset) = 0$.
3) $\varphi(A \cup B) \leq \varphi(A) + \varphi(B)$.

The submeasure φ is called a Maharam submeasure if it is sequentially point continuous, that is, if $\varphi(A_n) \dashrightarrow \varphi(A)$ for all sequence $A_n \in \mathcal{B}$ with $\chi_{A_n} \dashrightarrow \chi_A$ pointwise.

D. Maharam asked the question whether or not any non vanishing Maharam submeasure admits a countably additive non negative bounded measure with the same zero sets. This problem seems to be open at the time of the present writing (June 1977).

The submeasure φ is called pathological if it does not dominate any bounded finitely additive non negative measure.

In the above definition $\mathcal{B}$ could be any Boolean algebra of subsets of X.

The following theorem is basic for the investigations in the present paper.

Theorem 1: Let φ be a non vanishing pathological submeasure defined on a Boolean algebra $\mathcal{A}$ of subsets of X. Let u be a bounded non vanishing finitely additive non negative measure defined on $\mathcal{A}$. Then φ is not u continuous, that is, the condition

$$1)\quad \forall e > 0 \quad \exists d > 0 \quad \forall A \in \mathcal{A} : u(A) \leq d \Rightarrow \varphi(A) \leq e$$

is not fulfilled.

Furthermore u is not φ continuous, that is, the condition

$$2)\quad \forall e > 0 \quad \exists d > 0 \quad \forall A \in \mathcal{A} : \varphi(A) \leq d \Rightarrow u(A) \leq e$$

is not fulfilled.

Proof: It is of course no restriction to assume in the proof, as we shall do, that $\mathcal{A}$ separates points of X.

Let Ω be the compact Stone space of $\mathcal{A}$ (in which X is canonically imbedded). There is a bijective correspondence between sets $A \in \mathcal{A}$ and clopen sets $\tilde{A}$ in $\tilde{\mathcal{A}}$ (the clopen sets of Ω) given by

$$\tilde{A} \cap X = A .$$

Of course this correspondence is a Boolean algebra isomorphism. The submeasure φ is lifted, using this correspondence, to a pathological submeasure $\tilde{\varphi}$ on $\tilde{\mathcal{A}}$ and this is extended to a countably subadditive submeasure defined on all subsets of Ω by the definition

$$\tilde{\varphi}(B) = \inf\{\Sigma_i \tilde{\varphi}(A_i) \mid \quad B \subseteq \cup_i A_i \ , \ A_i \in \tilde{\mathcal{A}} \}$$

where the infimum is taken over all countable coverings of B with clopen sets. Of course the extended submeasure equals the pointwise supremum of all countably subadditive extensions.

The measure u defines a non trivial Radon measure $\tilde{u}$ on Ω by

$$\tilde{u}(\tilde{A}) = u(A) .$$

Suppose now that condition 1) holds. On the Borel field of Ω we define the metric d by

$$d(A,B) = \tilde{u}(A \Delta B) .$$

From the regularity of $\tilde{u}$ we easily see that $\tilde{\mathcal{A}}$ is d-dense in the Borel field of Ω. Since $\tilde{\varphi}$ is d-uniformly continous on $\tilde{\mathcal{A}}$ it has a unique d-continuous extension ψ to the Borel field. Of course ψ is a countably subadditive submeasure extending $\tilde{\varphi}$ to the Borel field of Ω. Since $\psi \leq \tilde{\varphi}$ on the Borel field, the submeasure ψ is pathological. The condition 1) implies the same condition with ψ, u and the Borel field of Ω. But this shows that $\tilde{u}(A) = 0$ implies $\psi(A) = 0$ for Borel sets A in Ω. This contradicts theorem 2 of [2] hence condition 1) is not fulfilled.

Suppose condition 2) holds. Then $\tilde{\varphi}(A) = 0$ implies $\tilde{u}(A) = 0$ for Borel sets in Ω. This contradiction with theorem 2 of [2] finishes the proof of theorem 1.

Both parts of thoerem 1 follows immediately from the next theorem. However, we shall need theorem 1 in the proof of the next theorem.

Theorem 2: Let $\mathcal{A}$ be a Boolean algebra of subsets of X and φ a non vanishing pathological submeasure defined on $\mathcal{A}$. Let u be a finitely additive non negative bounded measure on $\mathcal{A}$. For any $e > 0$ there exists $A \in \mathcal{A}$ with

$$\varphi(A) \leq e \qquad \text{and} \qquad u(A) \geq u(X)-e .$$

Proof: Let the submeasure ψ be defined as

$$\psi(A) = \inf\{u(B) + \varphi(C) \mid A \subseteq B \cup C \ , \ B,C \in \mathcal{A}\} .$$

From theorem 1 (observing that $\psi \leq u$) we conclude that ψ vanishes since otherwise φ would not be pathological. But this is the content of theorem 2. This concludes the proof.

It follows easily from theorem 2 that if φ_1 and φ_2 are pathological submeasures, then so is $\varphi_1+\varphi_2$.

Let φ_i $(i \in I)$ be a family of submeasures. The submeasure is continuous (by definition) with respect to this family if the condition

(C) For all $e > 0$ there is a $d > 0$ and a finite set $I^* \subseteq I$ such that $(\forall i \in I^* : \varphi_i(A) \leq d)$ implies $\varphi(A) \leq e$.

It follows from theorem 2 that if a submeasure is continuous with respect to some family of pathological submeasures then it is itself pathological.

From this remark we easily conclude that the set of pathological submeasures is closed in the topology induced by the metric D defined by

$$D(\varphi,\psi) = \sup\{|\varphi(A)-\psi(A)| \mid A \in \mathcal{A}\} .$$

The next result shows that for Maharam submeasures it is enough to look for control measures dominated by the submeasure. It has been noted independently by Christoph Bandt from Greifswald (unpublished) and very probable by other people too.

Theorem 3: If a Maharam submeasure φ defined on the measurable subsets of a measurable space $(X,\mathcal{B})$ admits a countably additive bounded non negative measure with the same zero sets (control measure), then φ admits a control measure dominated by φ.

Proof: Let u_i $(i \in I)$ be a family of non negative measures, all different from zero, dominated by φ and pairwise orthogonal. We choose a maximal family of this type.

Lemma: The maximal family of the type described above is non empty and countable.

Proof of lemma: Since φ admits a control measure, φ is absolutely continuous with respect to this measure and therefore theorem 1 implies that any submeasure dominated by φ is non pathological. Hence the family is non empty.

Suppose the family is uncountable. Then there is a sequence u_{i_n} of different elements in the family and a suitable $e > 0$ such that $u_{i_n}(X) \geq e$ for all n. We choose then a sequence $B_n \in$ of pairwise disjoint sets such that $u_{i_n}(B_n) = u_{i_n}(X)$. Since $u_{i_n} \leq \varphi$ we have $\varphi(B_n) \geq e$. Since this contradicts the Maharam continuity condition, we conclude that u_i $(i \in I)$ is countable and the lemma is proved.

Now suppose that for some set $A \in \mathcal{B}$ we have $u_i(A) = 0$ for all $i \in I$ and $\varphi(A) > 0$. Then $\psi(B) = \varphi(A \cap B)$ is non pathological and therefore dominates a non trivial measure. But this contradicts the maximality of u_i $(i \in I)$. We conclude that for $A \in \mathcal{B}$

$$\varphi(A) = 0 \quad \Leftrightarrow \quad \forall_i : u_i(A) = 0 .$$

We may assume $I = N$ and from the above remarks it follows that

$$u = \sum_{n=1}^{\infty} 2^{-n} u_n$$

is a control measure dominated by φ. This finishes the proof.

Let $\mathcal{A}$ be a Boolean algebra of subsets of some set X (using the Stone representation the following could be formulated for an abstract Boolean algebra). A weak neighbourhood (of $\emptyset$) in $\mathcal{A}$ is by definition a subset of $\mathcal{A}$ containing a set of the form

$$N(e,F) = \{A \in \mathcal{A} \mid u(A) \leq e , u \in F \}$$

where $e > 0$ and F is a finite set of bounded finitely additive non negative measures on $\mathcal{A}$.

It turns out to be useful to have an intrinsic characterization of weak neighbourhoods.

<u>Theorem 4</u>: <u>A set</u> $S \subseteq \mathcal{A}$ <u>is a weak neighbourhood if and only if the following condition holds</u>:

<u>There is a</u> $d > 0$ <u>such that for all finite convex combinations</u> $c_i \geq 0$, $\Sigma_i c_i = 1$, $\Sigma_i c_i \chi_{A_i} \leq d$ <u>at least one</u>, A_{i_0} <u>say, of the sets</u> A_i <u>belongs to</u> S.

Proof: The condition is clearly necessary for being a weak neighbourhood. Suppose the condition is fulfilled, but S is not a weak neighbourhood. For each set (e,F) of an $e > 0$ and a finite set F of bounded finitely additive non negative measures on $\mathcal{A}$ we choose

$$A_{(e,F)} \in N(e,F) \setminus S .$$

The set of such pairs is ordered by

$$(e,F) \geq (d,G) \quad \Leftrightarrow \quad e \geq d \quad \text{and} \quad F \subseteq G .$$

We herewith obtain a net outside S tending to $\emptyset$ in the weak topology. Therefore

$$\{A_{(e,F)}\} = T$$

has $\emptyset$ as a weak accumulation point. The Hahn-Banach theorem now shows that the set of convex combinations of characteristic functions of sets in T has $\emptyset$ as a uniform accumulation point. The condition now gives a contradiction with $S \cap T = \emptyset$. This finishes the proof of theorem 4.

Theorem 5: Let φ be a pathological submeasure defined on a Boolean algebra $\mathcal{A}$ of subsets of X. Then for any $e > 0$ there is a convex combination $c_i \geq 0$, $\Sigma_i c_i = 1$, $\Sigma_i c_i \chi_{A_i} \geq 1-e$, $\varphi(A_i) \leq e$.

Proof: Let $H = \{A \in \mathcal{A} \mid \varphi(A) > e\}$. Then theorem 2 implies that $H^c = \{X \setminus A \mid A \in H\}$ is not a weak neighbourhood. Hence there exists a convex combination $c_i \geq 0$, $\Sigma c_i = 1$, $\Sigma c_i \chi_{B_i} < e$, $B_i \notin H^c$.

Then the sets $A_i = X \setminus B_i$ has the properties of the theorem and the proof is finished.

In the next theorem we consider the unit interval $I = [0,1]$ with normalized Lebesgue measure v .

Theorem 6: Let $(X, \mathcal{B})$ be a measurable space and φ a Maharam submeasure defined on $\mathcal{B}$. Then φ has a control measure if and only if the following condition holds:

For all measurable sets $A \subseteq X \times I$ the condition

$$\forall x \in X : v(\{y \in I \mid (x,y) \in A\}) = 0$$

implies the existence of a point $y_0 \in I$ with

$$\varphi(\{x \in X \mid (x,y_0) \in A\}) = 0 .$$

If the Maharam submeasure φ above is pathological then there is a measurable set $A \subseteq X \times I$ such that

$$\forall x \in X : v(\{y \in I \mid (x,y) \in A\}) = 1 \quad \text{and}$$
$$\forall y \in I : \varphi(\{x \in X \mid (x,y) \in A\}) = 0 .$$

Proof: Since every standard metric space with a diffuse probability measure is measure preserving Borel isomorphic to $I = [0,1]$, we could in the condition replace I with any compact metric space with some probability measure (the last statement in the theorem would of course only be true if the measure were diffuse).

We start the proof by showing that

$$S_{\varphi,e} = \{A \in \mathcal{B} \mid \varphi(A) < e\}$$

is a weak neighbourhood for every $e > 0$. Suppose this is not true for some $e > 0$.

According to theorem 4 we can choose a sequence

$$c_i^n > 0 \,, \quad \Sigma_{i=1}^{p_n} \; c_i^n = 1, \quad \Sigma_{i=1}^{p_n} \; c_i^n \chi_{A_{i_n}^n} \leq 2^{-n}$$

of finite convex combinations such that

$$A_{i_n}^n \notin S_{\varphi,e} \qquad \text{for } 1 \leq i \leq p_n$$

and the $A_{i_n}^n$'s are $\mathcal{B}$ sets. We consider now Ω, the compact metrizable space of all sequences i_n with $1 \leq i_n \leq p_n$ (equipped with the topology of pointwise convergence). We define on Ω the probability measure u as the product of the measures

$$u_n = \Sigma_{i=1}^{p_n} \; c_i^n \varepsilon_i \quad .$$

Now it is easily seen that for each particular $x \in X$ we have that for u-almost every sequence $1 \leq i_n \leq p_n$

$$(*) \qquad \lim_{n\to\infty} \chi_{A_{i_n}^n}(x) = 0$$

(unfortunately this set of sequences may depend on $x \in X$).

An application of the condition (assumed from the beginning to hold) shows that there is a sequence $1 \leq i_n \leq p_n$ such that this limit equals zero for φ-almost every x. But this implies $\varphi(A_{i_n}^n) \to 0$ and a contradiction is obtained. It follows that there exists a finitely additive probability measure u (convex combination of measures determining the $S_{\varphi,e}$'s) such that φ is u-continuous. An application of theorem 1 now shows that no submeasure dominated by φ is pathological and the existence of a control measure is now proved in a way almost identical to the proof of theorem 3. This finishes the proof of the first part of the theorem.

Suppose now that the measure φ is pathological. We choose a sequence of convex combinations according to theorem 5 with $e = 2^{-n}$ and define Ω and u as in the above proof. The set $A \subseteq X \times \Omega$ is the set of all pairs of $x \in X$ and sequences $i_n \in \Omega$ such that the limit in (*) equals 1. It is easy to see that the properties of the theorem holds. This finishes the proof.

The control measure problem (in the Maharam setup) seems to be open at the time of this writing (June 1977). It ought to be true that an invariant Maharam submeasure defined on the Borel sets of an abelian metrizable compact topological group has the same zero sets as the Haar measure. We have no complete proof of this conjecture. It should be interesting to know whether or not an integral of pathological submeasures is again pathological (ex. a convolution of a pathological submeasure with Haar measure).

REFERENCES:

1) Jens Peter Reus Christensen, Topology and Borel structure, Notas de matematica nr. 10, North-Holland.

2) J.P.R. Christensen & Wojchiech Herer, On the existence of pathological submeasures and the construction of exotic groups, Math. Ann. 213, 203-210 (1975).

3) Maharam D., An algebraic characterization of measure algebras, Ann. Math. 48, 154-157 (1947).

ON MEASURABLE AND PARTITIONABLE VECTOR VALUED MULTIFUNCTIONS

by

R. Delanghe and C. Blondia

1. Throughout this paper X is an abstract space, H a σ-algebra of subsets of X and H_0 a subfamily of H consisting of null sets, which means that

(i) if $\alpha \subset \beta \in H_0$ then $\alpha \in H_0$;

(ii) if $\alpha_n \in H_0$, $n \in N$, then $\bigcup_{n\in N} \alpha_n \in H_0$.

In [5] and [6] M. Sion has introduced the notion of partitionability for a multifunction $f : X \to E$, E being a uniform space. This concept, which has its origin in [4], plays a fundamental role in the theory of integration developed in the second chapter of [6].
In this paper we shall deal with partitionable multifunctions $f : X \to E$ where E is a complex vector space provided with a system of seminorms. Our main result in §2 concerns a characterization of partitionable multifunctions when E is a hyperstrict inductive limit of spaces E_i each of which is provided with a countable system of seminorms. Using this characterization, which in fact generalizes a result given by M. Sion [6], we have been able to obtain some connections between partitionable and various types of vector valued measurable functions (§3,§4).

2. Let E be a uniform space with uniformity U and $f : X \to E$ be a multifunction. Then for $A \subset X$ and $F \subset E$ we call

$$f[A] = \bigcup_{x\in A} f(x)$$

and

$$f_{-1}[F] = \{x\in X : f(x)\subset F\}.$$

<u>Definitions 1</u>. ([6], II.2.4.) If $f : X \to E$ is a multifunction, then
(i) f is almost single-valued if there exists $X_0 \in H_0$ such that

$f|X \setminus X_0$ is single-valued ;

(ii) f is almost separably-valued if there exists $X_0 \in H_0$ such that $f[X \setminus X_0]$ is separable ;

(iii) f is quasi-bounded if for every $u \in U$, there exist $X_0 \in H_0$ and a countable subset F of E such that $f[X \setminus X_0] \subset u[F]$.

(iv) f is partitionable if for every $u \in U$, there exists a countable, disjoint $P \subset H$ such that $X \setminus (\bigcup_{\alpha \in P} \alpha) \in H_0$ and for every $\alpha \in P$, $f[\alpha] \times f[\alpha] \subset u$.

For the terminology used in the sequel concerning vector spaces provided with a system of seminorms, we refer to [1].

The properties mentioned in the following lemma may be easily checked and they are therefore given without proof.

Lemma 1. (i) Let E be a vector space, p be a seminorm on E and B and A be subsets of E such that B is countable and dense for p in A. Then A is separable for p.

(ii) Let either E be a vector space provided with a countable system of seminorms P or a strict inductive limit of such spaces. Let furthermore B and A be subsets of E such that B is countable and dense in A. Then A is separable.

Proposition 1. Let E be a vector space provided with a system of seminorms P and $f : X \to E$ be a multifunction. Then f is quasi-bounded if and only if for each $p \in P$, there exists $X_{0,p} \in H_0$ such that $f[X \setminus X_{0,p}]$ is separable for p.

Proof. The sufficient condition being trivial, we pass to the proof of the necessary condition.

Take $p \in P$ fixed and consider for each $n \in N$, $u_{p,n} \in U(P)$ defined by

$$u_{p,n} = \{(x,y) \in E^2 : p(x-y) \leq \frac{1}{n}\} .$$

As f is quasi-bounded, there exist a countable set $F_n \subset E$ and $X_{0,n} \in H_0$ such that $f[X \setminus X_{0,n}] \subset u_{p,n}[F_n]$.

Call $X_0 = \bigcup_{n\in\mathbb{N}} X_{0,n}$ and $F = \bigcup_{n\in\mathbb{N}} F_n$; then $X_0 \in \mathcal{H}_0$ and $F \subset E$ is countable. Moreover, if $y \in f[X\setminus X_0]$, then $y \in f[X\setminus X_{0,n}]$ for each $n \in \mathbb{N}$ so that $y_n \in F_n \subset F$ may be found for which $p(y-y_n) \leq \frac{1}{n}$.
Hence F is dense for p in $f[X\setminus X_0]$ so that, in view of Lemma 1(i), $f[X\setminus X_0]$ is separable for p. □

It is clear from the definition that an almost separably-valued function is quasi-bounded. A partial converse, sufficient for our further needs, is given in the following lemma.

Lemma 2. Let E be a vector space provided with a countable system of seminorms $\mathcal{P}$ and $f : X \to E$ be quasi-bounded. Then f is almost separably-valued.
Proof. Let $p \in \mathcal{P}$; then there exists $X_{0,p} \in \mathcal{H}_0$ such that $f[X\setminus X_{0,p}]$ is separable for p.
Call $X_0 = \bigcup_{p\in\mathcal{P}} X_{0,p}$; then $X_0 \in \mathcal{H}_0$ and $f[X\setminus X_0] \subset f[X\setminus X_{0,p}]$ for each $p \in \mathcal{P}$. Since $\mathcal{P}$ is countable, $f[X\setminus X_0]$ is separable from which it follows that f is almost separably-valued. □

In [6], II.2.7. M. Sion has proved that if X_1 is a metric space and $f : X \to X_1$ is a multifunction, then f is partitionable if and only if

(i) f is almost single-valued

(ii) f is almost separably-valued

(iii) for every closed $F \subset X_1$, $f_{-1}[F] \in \mathcal{H}$.

From this result we may immediately derive that if E is a vector space provided with a countable system of seminorms $\mathcal{P}$, then the above conditions (i),(ii),(iii) are necessary and sufficient for a multifunction $f : X \to E$ to be partitionable.
We now wish to prove that a similar result remains valid in a more general situation.

Theorem 1. Let $(E,\mathcal{P})$ be a strict inductive limit of spaces $(E_i,\mathcal{P}_i)$ each of which is provided with a countable system of seminorms. Let furthermore $f : X \to E$ be a multifunction such that

(i) f is almost single-valued

(ii) f is almost separably-valued

(iii) $f_{-1}[F] \in H$ for every closed $F \subset E$.

Then f is partitionable.

Proof. Let $X_0^{\,1}, X_0^{\,2} \in H_0$ be such that $f|X\setminus X_0^{\,1}$ is single-valued and $f[X\setminus X_0^{\,2}]$ is separable and put $X_0 = X_0^{\,1} \cup X_0^{\,2}$. Then of course $X_0 \in H_0$ and $f|X\setminus X_0$ is single-valued while in view of Lemma 1, $f[X\setminus X_0]$ is separable.

Call for each $n \in N$ and $\pi \in P$,

$$u_{\pi,n} = \{(x,y)\in E^2 : \pi(x-y) \leq \tfrac{1}{n}\}$$

and let $D = \{y_k : k\in N\}$ be a countable dense subset of $f[X\setminus X_0]$. Let furthermore $B_k = b_\pi(y_k,\frac{1}{2n})$ be a closed π-semiball, $\alpha_k' = f_{-1}[B_k] \cap X\setminus X_0$ and put $\alpha_0 = \alpha_0'$, $\alpha_k = \alpha_k' \setminus \bigcup_{i=0}^{k-1} \alpha_i$, $k \geq 1$. Then $P = \{\alpha_k : k\in N\}$ is a countable disjoint subfamily of H.

If $x \in X\setminus X_0$, then $f(x) \in f[X\setminus X_0]$ so that $f(x) \in B_k$ for some k. Consequently $x \in f_{-1}[B_k] \cap X\setminus X_0 = \alpha_k'$ so that $x \in \bigcup_{k\in N} \alpha_k' = \bigcup_{k\in N} \alpha_k$. Hence $X\setminus X_0 \subset \bigcup_{k\in N} \alpha_k'$ and as by definition $\bigcup_{k\in N} \alpha_k' \subset X\setminus X_0$, we find that $X\setminus X_0 = \bigcup_{k\in N} \alpha_k$.

Let now $(y_1,y_2) \in f[\alpha_k]\times f[\alpha_k]$; then $(y_1,y_2) \in B_k$ so that $\pi(y_1-y_2) \leq \frac{1}{n}$ or $(y_1,y_2) \in u_{\pi,n}$.

As $\pi \in P$ and $n \in N$ have been taken arbitrarily, we may conclude that f is partitionable. □

Theorem 2. Let (E,P) be a hyperstrict inductive limit of spaces (E_i,P_i) each of which is provided with a countable system of seminorms. Let furthermore $f : X \to E$ be a partitionable multifunction.

Then (i) f is almost single-valued

(ii) f is almost separably-valued

(iii) $f_{-1}[F] \in H$ for all closed $F \subset E$.

Proof. We first prove that f is almost single-valued.

As $P_i = \{p_{i,n} : n\in N\} \leq P|E_i$, there exists for each $i,n \in N$ a seminorm $\pi_{i,n} \in P$ and a constant $c_{i,n} > 0$ such that

$$p_{i,n}(y) \leq c_{i,n}\pi_{i,n}(y) \quad , \quad y \in E_i \ .$$

Call $u_{i,n} = \{(x,y)\in E^2 : \pi_{i,n}(x-y) \leq \frac{1}{nc_{i,n}}\}$; then, f being partitionable, there exists a countable disjoint subfamily $P_{i,n}$ of H such that $X_{0,i,n} = X\setminus(\bigcup_{\alpha\in P_{i,n}} \alpha) \in H_0$ and for each $\alpha \in P_{i,n}$ $f[\alpha]\times f[\alpha] \subset u_{i,n}$.

Put $X_0 = \bigcup_{i,n\in N} X_{0,i,n}$; then $X_0 \in H_0$. Moreover, for each $x \in X\setminus X_0$, $f(x)$ is a singleton.

Indeed, suppose that for some $x \in X\setminus X_0$, $f(x)$ is not a singleton and take $y_1,y_2 \in f(x)$, $y_1 \neq y_2$. Then there exists $k \in N$ such that $y_1,y_2 \in E_k$ and since $x \in X\setminus X_0$, $x \in X\setminus X_{0,k,n}$ for all $n \in N$.

But for each $n \in N$ we may find just one $\alpha_{k,n} \in P_{k,n}$ such that $x \in \alpha_{k,n}$.

Hence $(y_1,y_2) \in f(x)\times f(x) \subset f[\alpha_{k,n}]\times f[\alpha_{k,n}] \subset u_{k,n}$ or

$\pi_{k,n}(y_1-y_2) \leq \frac{1}{nc_{k,n}}$.

Consequently, $p_{k,n}(y_1-y_2) \leq \frac{1}{n}$, $n \in N$, so that $y_1 = y_2$.

We now prove that for all $F \subset E$ closed, $f_{-1}[F] \in H$.

Let again $\pi_{k,n}$ and $c_{k,n} > 0$ be such that $p_{k,n}(y) \leq c_{k,n}\pi_{k,n}(y)$, $y \in E_k$, $k,n \in N$, and call for each $i \in N$, $F_i = F\cap E_i$.

Let furthermore for all $k,n \in N$,

$$u_{k,n} = \{(x,y)\in E^2 : \pi_{k,n}(x-y) \leq \frac{1}{nc_{k,n}}\} \ ;$$

then we prove that $\bigcap_{k,n\in N} u_{k,n}[F_i] = F_i$.

Clearly $F_i \subset \bigcap_{k,n\in N} u_{k,n}[F_i]$. Now take $y \in \bigcap_{k,n\in N} u_{k,n}[F_i]$ and choose $j \in N$ such that $y \in E_j$. Then for each $k,n \in N$, there exists $y_{k,n} \in F_i$ for which

$$\pi_{k,n}(y-y_{k,n}) \leq \frac{1}{nc_{k,n}} \ .$$

Fix $k \geq \max(i,j)$ and consider the sequence $(y_{k,m})_{m\in N}$; then

$y_{k,m} \underset{P}{\to} y$.

Indeed, let $\pi \in P$ be given such that π is associated with the respective sequences of seminorms and scalars $(p_{i,n_i})_{i\in N}$ and $(c_i)_{i\in N}$. Then, as $y \in E_k$ and $y_{k,m} \in F_i \subset E_k$, $\pi(y-y_{k,m}) \leqslant c_k p_{k,n_k}(y-y_{k,m})$, so that for $m \geqslant n_k$

$$\begin{aligned}\pi(y-y_{k,m}) &\leqslant c_k p_{k,m}(y-y_{k,m}) \\ &\leqslant c_k c_{k,m} \pi_{k,m}(y-y_{k,m}) \\ &\leqslant \frac{c_k}{m} .\end{aligned}$$

Hence $\lim\limits_{m\to\infty} \pi(y-y_{k,m}) = 0$.

As $\pi \in P$ has been taken arbitrarily and F_i is closed in E, $y \in F_i$ and so $\bigcap\limits_{k,n\in N} u_{k,n}[F_i] \subset F_i$.

In view of [6], II.2.5., we have that $f_{-1}[F_i] \in H$.

But by means of the first part of the proof, there exists $X_0 \in H_0$ such that $f|X\backslash X_0$ is single-valued. Hence there exist subsets A and A_i, $i \in N$, of X_0 for which

$$f_{-1}[F] = \{x\in X\backslash X_0 : f(x)\in F\} \cup A ,$$
$$f_{-1}[F_i] = \{x\in X\backslash X_0 : f(x)\in F_i\} \cup A_i .$$

We thus obtain that

$$\{x\in X\backslash X_0 : f(x)\in F\} = \bigcup_{i=1}^{\infty} \{x\in X\backslash X_0 : f(x)\in F_i\} \in H ,$$

from which it follows that $f_{-1}[F] \in H$.

Finally, to prove (ii),we first note that in view of the previous step $f_{-1}[E_i] \in H$ for each $i \in N$ and that by (i), there exists $X_0 \in H_0$ such that $f|X\backslash X_0$ is single-valued.

Call $X_i = \{x\in X\backslash X_0 : f(x)\in E_i\}$ and define $f_i : X_i \to E_i$ by $f_i(x) = f(x)$, $i \in N$. Then $X_i \in H$ since $f_{-1}[E_i] = \{x\in X\backslash X_0 : f(x)\in E_i\} \cup A_i$, where $A_i \in H_0$.

Put $\mathcal{H}_i = X_i \cap \mathcal{H} = \{X_i \cap H : H \in \mathcal{H}\}$ and $\mathcal{H}_{0,i} = X_i \cap \mathcal{H}_0 = \{X_i \cap H_0 : H_0 \in \mathcal{H}_0\}$; then we prove that for each $i \in \mathbb{N}$, $f_i : X_i \to E_i$ is quasi-bounded.
Indeed, let for all $i,n \in \mathbb{N}$, $\pi_{i,n} \in \mathcal{P}$ and $c_{i,n} > 0$ be such that

$$p_{i,n}(y) \leq c_{i,n}\pi_{i,n}(y) \quad , \quad y \in E_i \ .$$

As f is partitionable, f is quasi-bounded ([6], II.2.6.). We may so find $X_{0,\pi_{i,n}} \in \mathcal{H}_0$ and a countable subset $F_{\pi_{i,n}}$ of E which is dense for $\pi_{i,n}$ in $f[X \setminus X_{0,\pi_{i,n}}]$.
Put $X_{0,i,n} = X_{0,\pi_{i,n}} \cap X_i \in \mathcal{H}_{0,i}$; then, since $f_i[X_i \setminus X_{0,i,n}] \subset f[X \setminus X_{0,\pi_{i,n}}]$, $F_{\pi_{i,n}}$ is dense for $\pi_{i,n}$ in $f_i[X_i \setminus X_{0,i,n}]$ so that, in view of Lemma 1(i), $f_i[X_i \setminus X_{0,i,n}]$ is separable for $\pi_{i,n}$. But as the latter set is contained in E_i and in E_i, $p_{i,n} \leq \pi_{i,n}$, we obtain that $f_i[X_i \setminus X_{0,i,n}]$ is separable for $p_{i,n}$.
Consequently, f_i is quasi-bounded and this implies, in virtue of Proposition 1, that f_i is almost separably-valued.
Hence, for each $i \in \mathbb{N}$, there exists $X_{0,i} \in \mathcal{H}_{0,i}$ such that $f_i[X_i \setminus X_{0,i}]$ is separable for $\mathcal{P}_i$.
Set $X_0' = \bigcup_{i=1}^{\infty} X_{0,i}$ and $X_0'' = X_0 \cup X_0'$; then $X_0', X_0'' \in \mathcal{H}_0$ and one may easily check that $f[X \setminus X_0''] = \bigcup_{i \in \mathbb{N}} f_i[X_i \setminus X_0'']$.
But as for any $i \in \mathbb{N}$, $f_i[X_i \setminus X_0''] \subset f_i[X_i \setminus X_{0,i}]$, we have that $f_i[X_i \setminus X_0'']$ is separable for $\mathcal{P}_i$.
Let therefore D_i be a countable dense subset of $f_i[X_i \setminus X_0'']$ and put $D = \bigcup_{i \in \mathbb{N}} D_i$; then clearly D is a countable dense subset of $f[X \setminus X_0'']$.
Consequently, $f[X \setminus X_0'']$ is separable for $\mathcal{P}$ which implies that f is almost separably-valued. □

Remarks. (1) It follows from Theorem 1 and 2 that if $(E,\mathcal{P})$ is a hyperstrict inductive limit of spaces $(E_i,\mathcal{P}_i)$ each of which is provided with a countable system of seminorms and if furthermore $f : X \to E$ is a multifunction, then f is partitionable if and only if the conditions (i), (ii) and (iii) are satisfied.

(2) Under the same assumptions of the previous remark, we obtain that a function $f : X \to E$ is partitionable if and only if

(i) f is almost separably-valued

(ii) f is measurable.

Here "f measurable" means that $f_{-1}[F] \in H$ for any closed subset F of E.

3. Let again X, H and H_0 be as in §1 and E be a complex vector space provided with a system of seminorms P.
In this section we deal with strongly measurable vector valued functions. The definition we give does not entirely correspond to the one used in the case of Banach spaces (see e.g. [3]) but in this case it coincides with the notion of Bochner measurability ([6], II.2.2.(2)). Let us first recall some classical notions :

(1) $g : X \to E$ is a simple function if range g is finite and for every $y \in$ range g, $g_{-1}[\{y\}] \in H$;

(2) $f : X \to C$ is measurable if for every closed $F \subset C$, $f_{-1}[F] \in H$;

(3) $f : X \to E$ is weakly measurable if for any bounded linear functional T on E, $T \circ f$ is measurable.

Definition 2. A function $f : X \to E$ is said to be strongly measurable if there exists $X_0 \in H_0$ and a sequence $(g_n)_{n \in N}$ of simple functions such that $g_n(x) \underset{P}{\to} f(x)$ for all $x \in X \setminus X_0$.

In [6] II.2.8.(1) M. Sion has shown that if E is a Banach space, then $f : X \to E$ is partitionable if and only if f is strongly measurable. If one analyses carefully his proof of the sufficient condition, then one may observe that the same line of thought can be repeated if instead of working with a norm, a metric or a semi-metric is used. We so arrive at

Theorem 3. If E is a vector space provided with a countable system of seminorms $\mathcal{P}$, then a function $f : X \to E$ is partitionable if and only if f is strongly measurable.

<u>Proof</u>. Let p_m, $m \in \mathbb{N}$, be an arbitrary element of $\mathcal{P}$ and define for each $x,y \in E$,

$$d(x,y) = \sum_{m=1}^{\infty} \frac{1}{2^m} \frac{p_m(x-y)}{1+p_m(x-y)} ;$$

then it is well known that d is a metric on E whose topology is equivalent to the one induced by $\mathcal{P}$.

Using this metric d, the proof of the necessary condition may then be carried out as in [6],II.2.8.

Conversely, suppose that f is strongly measurable and let $X_0 \in \mathcal{H}_0$ and $(g_n)_{n\in\mathbb{N}}$ be a sequence of simple functions such that $g_n(x) \underset{\mathcal{P}}{\to} f(x)$ for all $x \in X\setminus X_0$. Then for each $n \in \mathbb{N}$ and $T \in E^*$, $T\circ g_n$ is a simple function on $X\setminus X_0$. Moreover, as on $X\setminus X_0$, $T\circ f(x) = \lim_{n\to\infty} T\circ g_n(x)$, $T\circ f$ is measurable on it. Since $T\circ f$ is clearly measurable on X_0, we so obtain that $T\circ f$ is measurable on X, from which it follows that f is weakly measurable.

We now prove that f is quasi-bounded.

Indeed, since $(g_n)_{n\in\mathbb{N}}$ is a sequence of simple functions which converges pointwise to f on $X\setminus X_0$, $G = \bigcup_{n\in\mathbb{N}} g_n[X]$ is a countable subset of E which is dense in $f[X\setminus X_0]$ and so, in view of Lemma 1, $f[X\setminus X_0]$ is separable.

As f is weakly measurable and quasi-bounded, f is partitionable ([6],II.3.6.). □

We now wish to extend the foregoing result to the case where $(E,\mathcal{P})$ is a hyperstrict inductive limit of spaces $(E_i,\mathcal{P}_i)$ each of which is equipped with a countable system of seminorms.

First remark that, in virtue of Theorem 2, we then have that for a partitionable function $f : X \to E$, $f_{-1}[E_i] \in \mathcal{H}$ for all $i \in \mathbb{N}$.

If we put for any $i \in \mathbb{N}$, $X_i = f_{-1}[E_i]$, $\mathcal{H}_i = X_i\cap\mathcal{H} = \{X_i\cap H : H\in\mathcal{H}\}$ and $\mathcal{H}_{0,i} = X_i\cap\mathcal{H}_0 = \{X_i\cap H_0 : H_0\in\mathcal{H}_0\}$, then we define $f_i : X_i \to E$ by $f_i = f|X_i$.

Of course only those $i \in \mathcal{N}$ should be considered for which $X_i \neq \phi$, but as $E_i \subset E_{i+1}$, we may assume without loss of generality that $f[X] \cap E_1 \neq \phi$.

Lemma 3. Let $(E,\mathcal{P})$ be a hyperstrict inductive limit of spaces $(E_i,\mathcal{P}_i)$ each of which is provided with a countable system of seminorms and let $f : X \to E$ be partitionable. Then $f_i : X_i \to E_i$ is partitionable for each $i \in \mathcal{N}$.

Proof. In view of [6],II.2.7. it suffices to verify that f_i is almost separably-valued and that $(f_i)_{-1}[F] \in \mathcal{H}_i$ for each closed $F \subset E_i$.
As f is partitionable we have in view of Theorem 2 that f is almost separably-valued and so there exists $X_0 \in \mathcal{H}_0$ such that $f[X \setminus X_0]$ is separable.
Call $X_{0,i} = X_i \cap X_0$; then $X_{0,i} \in \mathcal{H}_{0,i}$ and as $f_i[X_i \setminus X_{0,i}] \subset f[X \setminus X_0] \cap E_i$, $f_i[X_i \setminus X_{0,i}]$ is separable for $\mathcal{P}_i$ which implies that f_i is almost separably-valued.
Finally, let $F \subset E_i$ be closed ; then F is closed in E so that in view of Theorem 2, $(f_i)_{-1}[F] = f_{-1}[F] \in \mathcal{H} \cap X_i = \mathcal{H}_i$. □

Corollary. Let $(E,\mathcal{P})$ and $f : X \to E$ be as in Lemma 3. Then for each $i \in \mathcal{N}$, $f_i : X_i \to E_i$ is strongly measurable.
Proof. Apply Theorem 3. □

Theorem 4. Let $(E,\mathcal{P})$ be a hyperstrict inductive limit of spaces $(E_i,\mathcal{P}_i)$ each of which is equipped with a countable system of seminorms. Then a function $f : X \to E$ is partitionable if and only if f is strongly measurable.
Proof. If f is strongly measurable then similar arguments as used in Theorem 3 yield the partitionability of f.
Suppose now that f is partitionable ; then from the above corollary , it follows that f_i is strongly measurable for each $i \in \mathcal{N}$. But this implies that $X_{0,i} \in \mathcal{H}_{0,i}$ and sequences $(g_n^i)_{n\in\mathcal{N}}$ of simple functions on X_i may be found such that $g_n^i(x) \xrightarrow{\mathcal{P}_i} f_i(x)$, $x \in X_i \setminus X_{0,i}$.

Define for each $i \in N$,

$$g_{i,n}(x) = \begin{cases} g_n^i(x) & , x \in X_i \\ 0 & , x \in X \backslash X_i \end{cases}$$

and $\quad f_i' = f\delta_{X_i}$.

Then $(g_{i,n})_{n\in N}$ is a sequence of simple functions on X such that $g_{i,n}(x) \underset{P}{\to} f_i'(x)$ for all $x \in X\backslash X_{0,i}$ and so f_i' is a strongly measurable function.

Call $Y_0 = \bigcup_{i\in N} X_{0,i}$ and define for any $M \in N$ and $x \in X$, with $X_0 = \phi$,

$$g_M(x) = \sum_{i=1}^{M} g_{i,M}(x)\,\delta_{X_i\backslash X_{i-1}}(x) .$$

Then $(g_M)_{M\in N}$ is a sequence of simple functions on X such that $g_M(x) \underset{P}{\to} f(x)$ for all $x \in X\backslash Y_0$.

Indeed, take $\pi \in P$ and $x \in X\backslash Y_0$ fixed. Then there exists $k \in N$ for which $x \in X_k\backslash X_{k-1}$ and $x \in X_k\backslash X_{0,k}$.

For all $M \geqslant k$ we so obtain that $g_M(x) = g_{k,M}(x)$.

But as $f(x) = f_k'(x)$ we find that, given $\varepsilon > 0$, there exists $M(\varepsilon) \in N$ such that $\pi(g_{k,m}(x) - f_k'(x)) \leqslant \varepsilon$ if $m \geqslant M(\varepsilon)$.

Hence $M \geqslant \max(M(\varepsilon),k)$ implies that $\pi(g_M(x) - f(x)) \leqslant \varepsilon$.

We may so conclude that f is strongly measurable. □

4. If Ω is an open subset of R^n, μ is a measure in Ω and E is a vector space provided with a system of seminorms P, then in [2] an integration theory has been built up for E-valued functions defined in Ω. A central role in this theory is played by the set of functions f which are μ-measurable by seminorm ; this means that for each $p \in P$ there exists a sequence $(\alpha_m^p)_{m\in N}$ of step functions (fonctions étagées) in E such that $p(f(x) - \alpha_m^p(x)) \to 0$ μ.a.e.

If we look at the properties of the sets M_μ and N_μ consisting respectively of the μ-measurable and μ-null sets, then, as may be expected, they form a σ-algebra of subsets of Ω and a subfamily of null sets in the sense of §1.

It thus seems natural to extend the notion of μ-measurability by seminorm to a more abstract setting and to look at its relationship with various types of measurability.

Let again X, $\mathcal{H}$, $\mathcal{H}_0$ and E be as in §1.

Definition 3. A function $f : X \to E$ is said to be measurable by seminorm if for each $p \in \mathcal{P}$ there exist $X_{0,p} \in \mathcal{H}_0$ and a sequence $(g_n^p)_{n\in\mathbb{N}}$ of simple functions on X such that for all $x \in X\setminus X_{0,p}$ $\lim_{n\to\infty} p(f(x)-g_n^p(x)) = 0$.

Theorem 5. If E is a vector space provided with a system of seminorms $\mathcal{P}$, then a function $f : X \to E$ is partitionable if and only if f is measurable by seminorm.

Proof. Assume that f is partitionable, take $p \in \mathcal{P}$ and consider the semimetric d_p defined by $d_p(x,y) = p(x-y)$, $x,y \in E$.

Then by means of an analogous reasoning as in [6],II.2.8., one may prove that f is measurable by seminorm.

Conversely, suppose that f is measurable by seminorm and choose $T \in E^*$ arbitrarily. Then there exists $p \in \mathcal{P}$ and $C > 0$ such that

$$|T(y)| \leq Cp(y) \quad , \quad y \in E .$$

But as f is measurable by seminorm, there exists $X_{0,p} \in \mathcal{H}_0$ and a sequence of simple functions $(g_n^p)_{n\in\mathbb{N}}$ such that for all $x \in X\setminus X_{0,p}$ $\lim_{n\to\infty} p(f(x)-g_n^p(x)) = 0$.

Hence $\lim_{n\to\infty} T\circ g_n^p(x) = T\circ f(x)$ for any $x \in X\setminus X_{0,p}$ which implies that $T\circ f$ is measurable on $X\setminus X_{0,p}$ so that, $T\circ f$ being measurable on $X_{0,p}$, we finally obtain that $T\circ f$ is measurable on X.

Consequently f is weakly measurable.

We now prove that f is quasi-bounded.

Take therefore $p \in \mathcal{P}$ and let $X_{0,p} \in \mathcal{H}_0$ and $(g_n^p)_{n\in\mathbb{N}}$ be a sequence of simple functions such that $\lim_{n\to\infty} (f(x)-g_n^p(x)) = 0$ for all $x \in X\setminus X_{0,p}$. Then clearly $D_p = \bigcup_{n\in\mathbb{N}} g_n^p[X]$ is a countable set which is dense for p in $f[X\setminus X_{0,p}]$ so that, in view of Lemma 1, $f[X\setminus X_{0,p}]$ is separable for p.

By means of [6], II.3.6. we may conclude that f is partitionable. □

Bringing together the results of this paper and [6],II.3.6., we have the following.

Theorem 6. Let either E be a vector space provided with a countable system of seminorms or a hyperstrict inductive limit of such spaces. Then for a function $f : X \to E$, the following statements are equivalent:

(i) f is partitionable

(ii) f is strongly measurable

(iii) f is weakly measurable and almost separably-valued

(iv) f is measurable by seminorm.

Note that [(ii)⇔(iii)] is a generalization of the classical B.J. Pettis theorem while [(ii)⇔(iv)] extends a result obtained in [2], p.247.

References

1. H.G. Garnir, M. De Wilde, J. Schmets, Analyse fonctionnelle, T. I, Théorie générale (Birkhäuser Verlag, Basel, 1968).
2. ——————, ——————, —————, Analyse fonctionnelle, T.II, Mesure et intégration dans l'espace euclidien (Birkhäuser Verlag, Basel, 1972).
3. E. Hille and R.S. Phillips, Functional analysis and semigroups (American Mathematical Society Colloquium Publications, Vol. XXXI, 1957).
4. C.E. Rickart, Integration in a convex linear topological space, Trans. Amer. Math. Soc. 52 (1942), 498-521.
5. M. Sion, Lectures on vector valued measures (University of British Columbia, 1969-70).
6. ————, A theory of semigroup valued measures (Lecture Notes in Mathematics 355, Springer Verlag, Berlin, 1973).

Seminar of Higher Analysis, State University of Ghent,
Krijgslaan 271, B-9000 Gent (Belgium).

Analytic Evolution Equations in Banach Spaces[1]

by

Thomas A. W. Dwyer, III

Northern Illinois University

0. Introduction. In this article we study infinite-dimensional versions of the equations

$$(\partial/\partial\tau)h(\tau,\xi) + \textstyle\sum_j f_j(\tau,\xi)(\partial/\partial\xi_j)h(\tau,\xi) = 0 \ , \tag{1}$$

where $h(\tau,\xi)$ is known when $\tau = t$,

$$(\partial/\partial t)\phi(x) = \textstyle\sum_j x_j f_j(t,\partial/\partial x_j)\phi(t,x) \ , \tag{2}$$

where $\phi(t,x)$ is known when $t = \tau$, and of the system

$$(\partial/\partial t)\xi_j(t) = f_j(\xi(t)) \ , \tag{3}$$

where $\xi(t) := (\xi_j(t))_j$ is known when $t = \tau$. We let t and τ be in $\mathbb{K} := \mathbb{R}$ or $\mathbb{C}$, $x := (x_j)_j$ in a vector space E over $\mathbb{K}$, $\xi := (\xi_j)_j$ in the dual space E' with respect to the form $\langle x,\xi\rangle := \sum_j x_j\xi_j$, the $f_j(t,\xi)$ being of class C^m (with $m \geq 0$) in a neighborhood of t and analytic in a neighborhood of ξ .

Equations (1) and (2) arise as linear realizations of nonlinear analytic dynamical systems evolving in E' in the following sense: given an analytic function $h(\xi)$, let $y(t) := h(\xi(t))$ be a nonlinear "output measurement at time t" of the system with state $\xi(\tau) = \xi$ when $t = \tau$ and governed by equation (3). Let also $\Phi(t,\tau)\phi(\tau,x) := \phi(t,x)$ be the resolvent operator of equation (2), and $H(\tau,t)h(t,\xi) := h(\tau,\xi)$ the resolvent operator of equation (1). Letting $h(\partial/\partial x)$ denote the partial differential operator of infinite order obtained when ξ_j is replaced by $\partial/\partial x_j$ in $h(\xi)$ for each j , it turns out that

$$h(\xi(t)) = \{h(\partial/\partial x)\Phi(t,\tau)\exp\langle x,\xi\rangle\}_{x=0} = H(\tau,t)h(\xi) \ .$$

That is, y(t) can also be regarded as a linear output measurement of the state $\phi(t,x)$ of the system governed by equation (2), evolving in a function space over E , or as a linear output measurement of the state $h(\tau,\xi)$ of the system governed by equation (1), evolving "backwards in time" in a function space over E' .

Research supported by NSF Grant MCS77-03900.

Approximate linear realizations of systems governed by equations of the form (3), in the tensor algebra of the original state space, where the new state vector is a direct sum of the tensor powers of $\xi(t)$, have been studied in [Brockett 1, 2], [Baillieul 1, 2] and elsewhere, and are intimately related to the function space realization given by equation (2), as described in [Dwyer 4]. Nonlinear systems evolving in sequence spaces appear when the space variables are discretized in the study of nonlinear parabolic equations by the method of lines, as in [Walter 1] and elsewhere. To include such situations, as well as stochastic systems as in [Ahmed 1] or distributed parameter systems when E is a Sobolev space, one must let E be infinite-dimensional. For this we first observe that equations (1), (2) and (3) can be written in coordinate-free form as

$$(\partial/\partial\tau)h_\tau(\xi) + \partial h_\tau(\xi)/\partial\vec{f}_\tau(\xi) = 0 , \qquad (1')$$

$$(\partial/\partial t)\phi_t(x) = \langle x,\vec{f}_t(\partial/\partial x)\phi_t(x)\rangle \qquad (2')$$

and

$$(\partial/\partial t)\xi_t = \vec{f}_t(\xi_t) , \qquad (3')$$

where $\vec{f}_t(\xi) := (f_j(t,\xi))_j$, $\partial/\partial\vec{f}_t(\xi)$ is the gradient operator in the direction of $\vec{f}_t(\xi)$, and $\vec{f}_t(\partial/\partial x)$ is the vector-valued partial differential operator of infinite order obtained when ξ_j is replaced by $\partial/\partial x_j$ in $f_t(\xi)$ for each j. For agreement with the notation of [Dwyer 2, 3], we will write

$$\vec{f}_\tau(\xi,d)h_\tau(\xi) := \partial h_\tau(\xi)/\partial\vec{f}_\tau(\xi) ,$$

$$h(d)\phi_t(x) := h(\partial/\partial x)\phi_t(x)$$

and

$$\vec{f}_t(d,x)\phi_t(x) := \langle x,\vec{f}_t(\partial/\partial x)\phi_t(x)\rangle$$

in the sequel . The transpose of $\vec{f}_\tau(\xi,d)$ is a hyperdifferential operator on E' , and $\vec{f}_t(d,x)$ is the Fourier-Borel transform of a hyperdiffernetial operator on E in the sense of [Treves 1], when K = C and E is finite-dimensional (and so, for that matter, are the resolvents of (1') and (2')). This suggests seeking estimates of "Ovcyannikov type" as in that reference, or in [Gelfand and Shilov 1, p.94] for K = R. In fact, the Ovcyannikov estimates for $f_\tau(\xi,d)$ (and $f_t(d,x)$ by duality), can be quickly obtained via Cauchy estimates as in the finite-dimensional case of [Treves 1, 2] when K = C: cf [Dwyer 4]. In sections 2 and 3 below we show that those estimates hold even when K = R for various analyticity types (Hilbert-Schmidt, nuclear, etc), but with respect to the norms of the function spaces introduced in section 1. They are then used to obtain uniqueness and iterative solutions to equations (1) and (2) in sections 4 and 5 respectively, and finally to obtain estimates of the output measurements of the system (3), given in section 6. As a final introductory remark we observe that the roles of E and E' can be interchanged, although it is convenient to have ξ_t in E' to handle the cases ξ_t in ℓ^∞ or L^∞ .

1. The spaces $\mathcal{F}^1_{\theta,\rho}$ and $\mathcal{F}^\infty_{\theta',\rho'}$. Let E be a Banach space over $\mathbb{K} = \mathbb{R}$ or $\mathbb{C}$, E' its dual; $\langle x,\xi\rangle := \xi(x)$ for x in E and ξ in E' ; $P(^nE';E')$ the space of continuous n-homogeneous polynomials $\vec{P}_n : E' \to E'$ with the norm $||\vec{P}_n|| :=$ $\sup\{||\vec{P}_n(x)||: ||x|| \leq 1\}$; $P_f(^nE';E')$ the subspace generated by the polynomials $v^n \otimes \nu : E' \to E'$ defined by $v^n \otimes \nu(\xi) := \langle v,\xi\rangle^n \nu$; $P_\theta(^nE';E')$ the completion of $P_f(^nE';E')$ for either: the nuclear norm induced by $E \otimes^n \otimes E'$ ($\theta = \mathbb{N}$); or the Hilbert-Schmidt norm if E is a Hilbert space ($\theta = H$); or the norm from $P(^nE';E')$ ($\theta = C$); or the norm L^p if $E = L^p$ and kernels are in L^p, $1 \leq p < \infty$: cf [Dwyer 2, 3, 4]. $P_\theta(^nE')$ is likewise defined in terms of polynomials $v^n : E' \to K$ given by $v^n(\xi) :=$ $\langle v,\xi\rangle^n$, so that the dual of $P_\theta(^nE')$ is the isometric image, under the map $\phi_n \mapsto \langle .,\phi_n\rangle_n$ characterized by $\langle v^n,\xi_n\rangle_n := \phi_n(v)$, of the space $P_{\theta'}(^nE)$ of n-homogeneous polynomials $\phi_n : E \to K$, respectively equal to: $P(^nE)$ if $\theta = N$; to $P_H(^nE)$ (Hilbert-Schmidt type on E) if $\theta = H$; to $P_I(^nE)$ (integral type on E) if $\theta = C$; or to the space of polynomials with kernels in $L^{p'}$ if $E = L^p$. Given also $\rho > 0$ let $\mathcal{F}^1_{\theta,\rho}(E';E')$ (resp. $\mathcal{F}^1_{\theta,\rho}(E')$) be the space of series $\vec{f} := \sum_{n=0}\vec{P}_n$ with $\vec{P}_n$ in $P_\theta(^nE';E')$ (resp. $f := \sum_{n=0}P_n$ with P_n in $P_\theta(^nE')$) such that $||\ \vec{f}\ ||_{\theta,\rho,1} :=$ $\sum_{n=0}\rho^n||\ \vec{P}_n||_\theta < \infty$ (likewise for $||\ f\ ||_{\theta,\rho,1}$): then each such $\vec{f}$ (resp. f) is analytic and $||\vec{f}(\xi)|| \leq\ ||\ \vec{f}\ ||_{\theta,\rho,1}$ if $||\xi|| \leq \rho$,with Fréchet derivative polynomials at the origin given by $\hat{d}^n\vec{f}(0) := n!\vec{P}_n$ (likewise for $|f(\xi)|$ and $\hat{d}^nf(0)$). Let also $\mathcal{F}^\infty_{\rho'}(E)$ be the space of series $\phi := \sum_{n=0}\phi_n$ with ϕ_n in $P_{\theta'}(^nE)$ such that $||\ \phi\ ||_{\theta',\rho',\infty} := \sup\{n \in \mathbb{N} : \rho'^n n!\,||\phi_n||_{\theta'}\} < \infty$: then each ϕ is an entire function and $|\phi(x)| \leq\ ||\ \phi\ ||_{\theta',\rho',\infty}\exp(1/\rho'\,||x||)$at each x in E, with Fréchet derivative polynomials at the origin given by $\hat{d}^n\phi(0) := n!\phi_n$. Finally, let

$$\langle\langle f,\phi\rangle\rangle := \sum_{n=0}(1/n!)\langle \hat{d}^nf(0),\hat{d}^n\phi(0)\rangle_n$$

(hence $\langle\langle f,\exp\langle .,\xi\rangle\ \rangle\rangle = f(\xi)$).

Proposition 1.1. $\mathcal{F}^1_{\theta,\rho}(E';E')$ (resp. $\mathcal{F}^1_{\theta,\rho}(E')$) and $\mathcal{F}^\infty_{\theta',\rho'}(E)$ are Banach spaces with the norms given by $||\ \vec{f}\ ||_{\theta,\rho,1}$ (resp. $||\ f\ ||_{\theta,\rho,1}$) and $||\ \phi\ ||_{\theta',\rho',\infty}$ respectively, and Taylor series converge to the corresponding functions for the norms given. Moreover, the map $\phi \mapsto \langle\langle .,\phi\rangle\rangle$ is an isometry from $\mathcal{F}^\infty_{\theta',\rho'}(E)$ onto the dual of $\mathcal{F}^1_{\theta,\rho}(E')$ if $\rho' = 1/\rho$.

Proof. Analogous to that for the spaces $\mathcal{F}^p_\theta(E')$ and $\mathcal{F}^{p'}_{\theta'}(E)$ in [Dwyer 1, Propositions 2.1.1, 2.1.2, and 2:1.3].

2. The operator $\vec{f}(.,d)$.

Definition 2.1. Given $\vec{f} : U \to E'$ defined on a neighborhood U of zero in E' and $h : U \to \mathbb{K}$, we write $[\vec{f}(.,d)h](\xi) := \vec{f}(\xi,d)h(\xi) := \langle \vec{f}(\xi), dh(\xi)\rangle$ (= $\langle dh(\xi), \vec{f}(\xi)\rangle$ when $dh(\xi)$ is in E) if the Fréchet differential $dh(\xi)$ is defined.

Proposition 2.1. Given $0 < \rho < \sigma$, if f is in $F^1_{\theta,\rho}(E';E')$ and h in $F^1_{\theta,\sigma}(E')$ then $\vec{f}(.,d)h$ is in $F^1_{\theta,\rho}(E')$, with

$$||\vec{f}(.,d)h||_{\theta,\rho,1} \leq ||h||_{\theta,\sigma,1}||\vec{f}||_{\theta,\rho,1}/(\sigma - \rho).$$

The proof consists of the successive application of lemmas 2.1, 2.2, 2.3 and 2.4 which follow.

Lemma 2.1. Given $\vec{P}_n$ in $P(^nE';E')$ and Q_{m+1} in $P(^{m+1}E')$, let $[\vec{P}_n(.,d)Q_{m+1}](\xi) := \langle\vec{P}_n(\xi,d)Q_{m+1}(\xi)\rangle := \langle\vec{P}_n(\xi), dQ_{m+1}(\xi)\rangle$ (= $\langle dQ_{m+1}(\xi), \vec{P}_n(\xi)\rangle$ since $dQ_{m+1}(\xi)$ is in E when Q_{m+1} is in $P_\theta(^{m+1}E')$): then $\vec{P}_n(.,d)Q_{m+1}$ is in $P_\theta(^{m+n}E')$, with

$$||\vec{P}_n(.,d)\ Q_{m+1}||_\theta \leq (m + 1)||Q_{m+1}||_\theta||\vec{P}_n||_\theta\ .$$

Proof. Follows from the identity

$$\vec{P}_n(\xi, dQ_{m+1}(\xi) = (m + 1)\langle Q_{m+1}, \xi^m.\vec{P}_n(\xi)\rangle_{m+1}$$

(where $\xi^m.\vec{P}_n(\xi)(x) := \langle x,\xi\rangle^m\langle x,\vec{P}_n(\xi)\rangle$), applied first to $Q_{m+1} = \langle v,.\rangle^{m+1}$ and $\vec{P}_n = \langle v,.\rangle^n \times \nu$, then extending by density to all Q_{m+1}, and finally to all P_n: cf [Dwyer 2, Proposition 1.2.1].

Lemma 2.2. Given $\vec{P}_n$ in $P(^nE';E')$ and $0 < \rho < \sigma$ let $h := Q_0 + \sum_{m=0} Q_{m+1}$ with $Q_k = (1/k!)d^kh(0)$ in $P_\theta(^kE')$: if h is in $F^1_{\theta,\sigma}(E')$ then

$$||\sum_{m=0}\vec{P}_n(.,d)Q_{m+1}||_{\theta,\rho,1} \leq ||h||_{\theta,\sigma,1}||\vec{P}_n||_\theta\rho^n/(\sigma - \rho)$$

for every M in $\mathbb{N}$.

Proof. The inequality of Lemma 2.1 yields

$$||\vec{P}_n(.,d)Q_{m+1}||_{\theta,\rho,1} \leq \rho^n||\vec{P}_n||_\theta(\,^m/m!)||d^{m+1}h(0)||_\theta\ .$$

If $a > 1$ then $1 < a^k/[k(a - 1)]$ for $k > 1$ (set $a = 1 + \varepsilon$ and use the binomial expansion), hence $m + 1 < a^{m+1}/(a - 1)$ for $m > 0$ (setting $m = k + 1$). Therefore

$$\rho^m/m! < \rho^m a^{m+1}/[(m + 1)!(a - 1)] = (a\rho)^{m+1}/[(m + 1)!(a\rho - \rho)] ,$$

so that $a := \sigma/\rho$ yields $\rho^m/m! < [\sigma^{m+1}/(m = 1)!]/(\sigma - \rho)$ .It is enough now to substitute this last inequality in the first estimate for $||P_n(.,d)Q_{m+1}||_{\Theta,\rho,1}$ and sum over m .

The next lemma follows immediately:

<u>Lemma</u> 2.3. Given $\vec{f} = \sum_{n=0}\vec{P}_n$ with $\vec{P}_n = (1/n!)d^n\vec{f}(0)$ in $P_\theta(^nE';E')$ and $\rho < \sigma$ let $\vec{f}$ be in $F^1_{\theta,\rho}(E';E')$. Keeping h and $Q_{m+1} = [1/(m + 1)!]d^{m+1}h(0)$ as in Lemma 2.2, let $L_n : E' \to \mathbb{K}$ be defined by

$$L_n(\xi) := \lim_M \sum_{m=0}^M \vec{P}_n(\xi,d)Q_{m+1}(\xi) :$$

then L_n is in $F^1_{\theta,\rho}(E')$, with

$$||\sum_{n=0}^N L_n||_{\theta,\rho,1} \leq ||h||_{\theta,\sigma,1}||\vec{f}||_{\theta,\rho,1}/(\sigma - \rho)$$

for each N in $\mathbb{N}$.

<u>Lemma</u> 2.4. With $\vec{f}$, h and L_n as above we have

$$<\vec{f}(\xi),dh(\xi)> = \lim_N \sum_{n=0}^N L_n(\xi)$$

(or $<dh(\xi),\vec{f}(\xi)>$, since $dh(\xi)$ is in E when h is of type θ in the sense of section 1).

<u>Proof</u>. Let $\vec{P}_n := (1/n!)d^n\vec{f}(0)$, so that

$$<dh(\xi),\vec{f}(\xi)> = \sum_{n=0}^\infty <dh(\xi),\vec{P}_n(\xi)> :$$

from [Nachbin 1, p.29] with $Q_{m+1} := (1/(m + 1)!d^{m+1}h(0)$ we get

$$<dh(\xi),\nu> = \sum_{m=0}^\infty <dQ_{m+1}(\xi),\nu>$$

for any ν in E' , so by setting $\nu := \vec{P}_n(\xi)$ and using the identity $<dQ_{m+1}(\xi),\vec{P}_n(\xi)> = \vec{P}_n(\xi,d)Q_{m+1}(\xi)$ of Lemma 2.1 we obtain $dh(\xi),\vec{P}_n(\xi) = L_n(\xi)$. It is enough now to sum over n.

<u>3. The operator</u> $\vec{f}(d,.)$.

Definition 3.1. Given $\vec{f} : U \to E'$ on a neighborhood U of zero in E' and $\phi : E \to \mathbb{K}$, let $\vec{f}_x(\xi) := \langle x, \vec{f}(\xi)\rangle$ and $T_{-x}\phi(v) := \phi(v + x)$ for x and v in E and ξ in E': if $\vec{f}_x$ is in $F^1_{\theta,\rho}(E')$ and $T_{-x}\phi$ in $F^{\infty}_{\theta',(1/\ )}(E)$ then we write $[\vec{f}(d,.)\phi](x) := \vec{f}(d,x)\phi(x) := \langle\langle \vec{f}_x, T_{-x}\phi\rangle\rangle$.

Proposition 3.1. Given $0 < \rho < \sigma$, let $\vec{f}$ be in $F^1_{\theta,\rho}(E';E')$ and ϕ in $F^{\infty}_{\theta',(1/\rho)}(E)$: then $\vec{f}(d,.)\phi$ is in $F^{\infty}_{\theta',(1/\sigma)}(E)$, with

$$||\vec{f}(d,.)\phi||_{\theta',(1/\sigma),\infty} \leq ||\phi||_{\theta',(1/\rho),\infty}||\vec{f}||_{\theta,\rho,1}\ \sigma/(\sigma - \rho)^2 .$$

The proof consists of the successive application of Lemmas 3.1, 3.2, 3.3 and 3.4 which follow.

Lemma 3.1. Given $\vec{P}_n$ in $P_\theta({}^nE';E')$ and ϕ_{m+n} in $P_{\theta'}({}^{m+n}E)$, if $\vec{P}_{nx}(\xi) := \langle x, \vec{P}_n(\xi)\rangle$ let $[\vec{P}_n(d,.)\phi_{m+m}](x) := \langle\vec{P}_n(d,x)\phi_{m+n}(x) := \langle \vec{P}_{nx}, \hat{d}^n\phi_{m+n}\rangle_n$: then

(a) $\vec{P}_n(d,.)\ {}_{m+n}$ is in $P_{\theta'}({}^{m+1}E)$, with

$$||\vec{P}_n(d,.)\phi_{m+n}||_{\theta'} < [(m + n)!/m!]\,||\vec{P}_n||_\theta||\phi_{m+n}||_{\theta'} ;$$

(b) If Q_{m+1} is in $P_\theta({}^{m+1}E')$ then

$$(m + 1)!\langle Q_{m+1}, \vec{P}_n(d,.)\phi_{m+n}\rangle_{m+1} := (m + n)!\langle\vec{P}_n(.,d)Q_{m+1}, \phi_{m+n}\rangle_{m+n} .$$

Proof. The identity (b) is first verified for $\vec{P}_n$ in $P_f({}^nE';E')$ and Q_{m+1} in $P_f({}^{m+1}E')$. This identity then yields the inequality (a) for polynomials in P_f by an argument using the Hahn-Banach and Alaoglu's theorems, as in the case of [Dwyer 2, Proposition 1.6.1], and is then extended by density to all Q_{m+1} . Finally, (a) and (b) are extended also by density to all $\vec{P}_n$.

Lemma 3.2. Given $\vec{P}_n$ in $P_\theta({}^nE';E')$ and $0 < \rho < \sigma$ let $\phi := \sum_{m=0}\phi_{m+n}$ with $\phi_{m+n} = [1/(m+n)!]\hat{d}^{m+n}\phi(0)$ in $P_{\theta'}({}^{m+n}E)$: if ϕ is in $F_{\theta',(1/\rho)}(E)$ then

$$||\sum_{m=0}^{M}\vec{P}_n(d,.)\phi_{m+n}||_{\theta',(1/\sigma),\infty} \leq \rho^n||\vec{P}_n||_\theta||\phi||_{\theta',(1/\rho),\infty}\sigma/(\sigma - \rho)^2$$

for every M in $\mathbb{N}$.

Proof. The inequality of Lemma 3.1 (a) (and the fact that $\vec{P}_n(d,.)\phi_{m+n}$ is (m +1)-homogeneous) yields

$$||\vec{P}_n(d,.)\phi_{m+n}||_{\theta',(1/\sigma),\infty} < (1/\sigma)^{m+1}(m + 1)\,||\vec{P}_n||_\theta||\hat{d}^{m+n}\phi(0)||_{\theta'} ,$$

from which one obtains

$$||\sum_{m=0}^{M}\vec{P}_n(d,.)\phi_{m+n}||_{\theta',(1/\sigma),\infty} \leq \rho^n||\vec{P}_n||_\theta||\phi||_{\theta',(1/\rho),\infty}(1/\sigma)\sum_{m=0}^{M}(m+1)(\rho/\sigma)^m .$$

It is enough now to observe that $\sum_{m=0}(m+1)t^m = 1/(1-t)^2$ for $|t| < 1$, and set $t = \rho/\sigma$.

The next lemma follows immediately:

Lemma 3.3. Given $\vec{f} = \sum_{n=0}^{\infty}\vec{P}_n$ with $\vec{P}_n = (1/n!)\hat{d}^n\vec{f}(0)$ in $P_\theta(^nE';E')$ and $\rho < \sigma$ let $\vec{f}$ be in $F^1_{\theta,\rho}(E';E')$. Keeping ϕ, $\phi_{m+n} = [1/(m+n)!]\hat{d}^{m+n}\phi(0)$ as in Lemma 3.2, let $\Lambda_n : E \to \mathbb{K}$ be defined by

$$\Lambda_n(x) := \lim_M\sum_{m=0}^{M}\vec{P}_n(d,x)\phi_{m+n}(x) :$$

then Λ_n is in $F^\infty_{\theta',(1/\sigma)}(E)$, with

$$||\sum_{n=0}^{N}\Lambda_n||_{\theta',(1/\sigma),\infty} \leq ||\phi||_{\theta',(1/\rho),\infty}||\vec{f}||_{\theta,\rho,1}\ \sigma/(\sigma-\rho)^2$$

for every N in $\mathbb{N}$.

Lemma 3.4. With f, ϕ, and Λ_n as above we have

$$<<\vec{f}_x,T_{-x}\phi>> = \lim_M\sum_{m=0}^{M}\Lambda_n(x) .$$

Proof. Clearly we have

$$<<\vec{f}_x,T_{-x}\phi>> = \sum_{n=0}^{\infty}<<\vec{f}_x,(1/n!)\hat{d}^n\phi(x)>> .$$

letting $\phi_{m+n} := [1/(m+n)!]\hat{d}^{m+n}\phi(0)$, from [Nachbin 1, p.30] we get

$$(1/n!)\hat{d}^n\phi(x) = \sum_{m=0}^{\infty}(1/n!)\hat{d}^n\phi_{m+n}(x) .$$

Letting $\vec{P}_n := (1/n!)\hat{d}^n\vec{f}(0)$ and using the identity $\vec{P}_n(d,x)\phi_{m+n}(x) = <\vec{P}_{nx},\hat{d}^n\phi_{m+n}(x)>_n$ of Lemma 3.1 we get

$$<<\vec{f}_x,(1/n!)\hat{d}^n\phi(0)>> = \sum_{m=0}^{\infty}<\vec{P}_{nx},\hat{d}^n\phi_{m+n}(x)>_n =: \Lambda_n(x) :$$

It is enough now to sum over n.

Proposition 3.2. Let $\vec{f}$ be in $F^1_{\theta,\rho}(E';E')$, h in $F^1_{\theta,\sigma}(E')$ and ϕ in $F^\infty_{\theta',(1/\rho)}(E)$: if $\rho < \sigma$ then $<<\vec{f}(.,d)h,\phi>> = <<h,\vec{f}(d,.)\phi>>$.

Proof. It is enough to set $\vec{P}_n := (1/n!)d^n\vec{f}(0)$, $Q_{m+1} := [1/(m+1)!]d^{m+1}h(0)$ and $\phi_{m+n} := [1/(m+n)!]d^{m+n}\,(0)$ in the identity of Lemma 3.1, observe that $\vec{P}_n(.,d)Q_{m+1}$ is (m+n)-homogeneous and $\vec{P}_n(d,.)\phi_{m+n}$ (m+1)-homogeneous and use the expansion of $\langle\langle .,.\rangle\rangle$ in terms of the bilinear forms $\langle .,.\rangle_n$, as in the "constant coefficients" case of [Dwyer], Proposition 1.9.1].

Since the norm of $\langle\langle .,\vec{f}(d,.)\phi\rangle\rangle$ in the dual of $F^1_{\theta,\sigma}(E')$ is $||\vec{f}(d,.)\phi||_{\theta',(1/\sigma),\infty}$ it folows from Propositions 3.1 and 3.2 that the estimate of Proposition 3.1 can be improved:

Corollary. With $\rho < \sigma$, $\vec{f}$ and ϕ as before, we have

$$||\vec{f}(d,.)\phi||_{\theta',(1/\sigma),\infty} \leq ||\phi||_{\theta',(1/\rho),\infty}||\vec{f}||_{\theta,\rho,1}/(\sigma - \rho) .$$

4. Evolution equations governed by $\vec{f}(.,d)$. In the sequel let $D_T(t) := \{\tau \in \mathbb{K} : |\tau - t| < T\}$; let also $C^n[D_T(t)]$ denote the space of n-times differentiable $\mathbb{K}$-valued functions on $D_T(t)$, and $C^n[D_T(t);X]$ the corresponding space of functions with values in a Banach space X.

Given $\sigma > 0$, $T > 0$ and $C > 0$, let there be given a family of functions $\vec{f}_\tau$ in $F^1_{\theta,\sigma}(E';E')$, such that $||\vec{f}||_{\theta,\sigma,1} < C$ for every τ in $D_T(t)$, and $\vec{f}_{(.)}$ is in $C^m[D_T(t);F^1_{\theta,\sigma}(E';E')]$ (where $\vec{f}_{(.)}(\tau) := \vec{f}_\tau$. In particular, if K = C and m > 0 then $\vec{f}_\tau$ is analytic in τ), such that

Theorem 4.1. Let $\vec{f}$ be as above. Given $0< \rho <\sigma$ and h in $F^1_{\theta,\sigma}(E')$ let $T < (\sigma - \rho)/Ce$: then there is a unique mapping $h_{(.)} : \tau \to h_\tau$ in $C^{m+1}[D_T(t);F^1_{\theta,\rho}(E')]$ (hence analytic in τ if K = C) such that

$$(\partial/\partial\tau)h_\tau(\xi) + \vec{f}_\tau(\xi, d)h_\tau(\xi) = 0$$

on $D_T(t)$ with $h_t = h$, given by $h\ (\xi) := \lim_k h_{k,\tau}(\xi)$ in $F^1_{\theta,\rho}(E')$, where $h_{0,\tau} := h$ and

$$h_{k,\tau}(\xi) := h(\xi) - \int_t^\tau \vec{f}_\lambda(\xi,d)h_{k-1,\lambda}(\xi)d\lambda$$

(straight line integral if $\mathbb{K} = \mathbb{C}$) for $k \geq 1$. The convergence is uniform on $D_{T'}(t)$ for T' > T, and

$$||h_\tau - h||_{\theta,\rho,1} \leq ||h||_{\theta,\sigma,1}Ce|\tau - t|/(\sigma - \rho - Ce|\tau - t|) .$$

Proof. From the estimate of Proposition 2.1 and the identity

$$h_{k,\tau}(\xi) - h_{k-1,\tau}(\xi) = \int_t^\tau f_\lambda(\xi,d)[h_{k-1,\lambda}(\xi) - h_{k-2,\lambda}(\xi)]d\lambda$$

we obtain

$$||h_{k,\tau} - h_{k-1,\tau}||_{\theta,\sigma-\varepsilon(1)-\ldots-\varepsilon(k),1} \leq ||h||_{\theta,\sigma,1}(C|\tau - t|)^k/k!\varepsilon(1)\ldots\varepsilon(k) .$$

The estimate for

$$\lim_k h_{k,\tau} = h + \sum_{k=1}^\infty (h_{k,\tau} - h_{k-1,\tau})$$

follows by setting $\varepsilon(1) = \ldots = \varepsilon(k) = (\sigma - \rho)/k$ and majorizing $k^k/k!$ by e^k. We get that h is of class C^{m+1} for $|\tau - t| < T' < T$ if $\vec{f}_\tau$ is of class C^m from the formula

$$h_\tau(\xi) = h(\xi) - \int_t^\tau \vec{f}_\lambda(\xi,d)h_\lambda(\xi)d\lambda ,$$

then letting T' tend to T. For uniqueness, if h_τ an $\tilde{h}_\tau$ are two C^1 solutions in $F^1_{\theta,\rho}(E')$ and $h_{t'} = \tilde{h}_{t'}$ for some t' then

$$h_\tau(\xi) - \tilde{h}_\tau(\xi) = \int_{t'}^\tau \vec{f}_\lambda(\xi,d)(h_\lambda - \tilde{h}_\lambda)(\xi)d\lambda ,$$

from which one derives

$$||h_\tau - \tilde{h}_\tau||_{\theta,\rho-\varepsilon(1)-\ldots-\varepsilon(k),1} \leq M(C|\tau - t|)^k/k!\varepsilon(1)\ldots\varepsilon(k)$$

with $\varepsilon(1) = \ldots = \varepsilon(k) = \varepsilon/k$ and $|\tau - t| < T' < T$, where $M := \max\{||h_\tau - \tilde{h}_\tau||_{\theta,\rho,1} : |\tau - t| \leq T'\}$. By letting k tend to ∞ we conclude that $h_\tau = \tilde{h}_\tau$ for $|\tau - t| < T'$, that is, the set S of points t' in $D_T(t)$ where $h_{t'} = \tilde{h}_{t'}$ is open. Since t is in S and S is also closed (by the continuity of $\tau \to h_\tau - \tilde{h}_\tau$), it follows that $h_\tau = \tilde{h}_\tau$ on $S = D_T(t)$.

In particular, let $\vec{f}_\tau = \vec{f}$ in $F^1_{\theta,\sigma}(E';E')$ for every τ in $D_T(t)$: we may then set $C := ||\vec{f}||_{\theta,\sigma,1}$ and write

$$\exp(\tau\vec{f}(.,d)) := \sum_{n=0}^\infty (\tau^n/n!)\vec{f}(.,d)^n$$

to get:

Corollary. Given $\vec{f}$ in $F^1_{\theta,\sigma}(E';E')$, h in $F^1_{\theta,\sigma}(E')$ and $\rho < \sigma$, let $T' < (\sigma - \rho)/e||\vec{f}||_{\theta,\sigma,1}$: then the solution $h_{(.)}$ of the equation in Theorem 4.1 with t = 0 is expressible as $h_\tau = \exp(-\tau\vec{f}(.,d))h$, hence $h_{(.)}$ is in $F^1_{T'}(\mathbb{K};F^1_{\theta,\rho}(E'))$, with $\{(\partial/\partial\tau)^n h\}_{\tau=0} = (-\vec{f}(.,d))^n h$ and

$$||h_{(.)}||_{T',1} \leq ||h||_{\theta,\sigma,1}(\sigma - \rho)/(\sigma - \rho - ||\vec{f}||_{\theta,\sigma,1}eT').$$

Proof. Follows from observing that now

$$h_{k,\tau}(\xi) = (-\vec{f}(.,d))^k h(\xi)\int_0^\tau\int_0^{\lambda_k}\ldots\int_0^{\lambda_2} d\lambda_1\ldots d\lambda_{k-1}d\lambda_k ,$$

then estimating $||\vec{f}(.,d)^k h||_{\theta,\sigma-\varepsilon(1)-\ldots-\varepsilon(k),1}$

5. Evolution equations governed by $\vec{f}(d,.)$. With the notation of section 4 we have

Theorem 5.1. Let $t \to \vec{f}_t$ be of class C^m in $D_T(\tau)$, with $||\vec{f}||_{\theta,\sigma,1} \leq C$ for t in $D_T(\tau)$. Given $0 < \rho < \sigma$ and ϕ in $F^\infty_{\theta',(1/\rho)}(E)$ let $T \leq (\sigma - \rho)/Ce$: then there is a unique mapping $\phi_{(.)} : t \to \phi_t$ in $C^{m+1}[D_T(\tau);F^\infty_{\theta',(1/\sigma)}(E)]$ (hence analytic in t if K = C) such that

$$(\partial/\partial t)\phi_t = \vec{f}_t(d,.)\phi_t$$

on $D_T(\tau)$ with $\phi_\tau = \phi$, given by $\phi_t(x) := \lim_k\phi_{k,t}(x)$ in $F^\infty_{\theta',(1/\ )}(E)$, where $\phi_{0,t} := \phi$ and

$$\phi_{k,t}(x) := \phi(x) + \int_\tau^t \vec{f}_\lambda(d,.)\phi_{k-1,\lambda}(x)d\lambda$$

for $k > 1$. The convergence is uniform on $D_{T'}(\tau)$ for $T' < T$, and

$$||\phi_t - \phi||_{\theta',(1/\sigma),\infty} \leq ||\phi||_{\theta',(1/\rho),\infty}Ce|t - \tau|/[\sigma - \rho - Ce|t - \tau|] .$$

Proof. From the corollary to Proposition 3.2 one derives

$$||\phi_{k,t} - \phi_{k-1,t}||_{\theta',(1/\sigma),\infty} \leq$$

$$\leq ||\phi||_{\theta',(1/\sigma-\varepsilon(1)-\ldots-\varepsilon(k)),\infty}(C|t - \tau|)^k/k!\varepsilon(1)\ldots\varepsilon(k)$$

and sets $\varepsilon(1) = \ldots = \varepsilon(k) = (\sigma - \rho)/k$ to estimate $||\phi_t||_{\theta',(1/\sigma),\infty}$. For uniqueness one lets k tend to ∞ in the similarly obtained estimate

$$||\phi_t - \tilde{\phi}_t||_{\theta',(1/\rho),\infty} \leq M(C|t - t|)^k/k!\varepsilon(1)\ldots\varepsilon(k)$$

for solutions ϕ_t and $\tilde{\phi}_t$ with $|t - t'| < T' < T$, such that $\phi_{t'} = \tilde{\phi}_{t'}$, $M := \max ||\phi_t - \tilde{\phi}_t||_{\theta',(1/\rho-\varepsilon(1)-\ldots-\varepsilon(k)),\infty} : |t - t'| \leq T'\}$ and $\varepsilon(1) = \ldots = \varepsilon(k) = \varepsilon/k$; cf proof of Theorem 4.1.

Setting $\tau = 0$ and $\vec{f}_t = \vec{f}$ in $F^1_{\theta,\sigma}(E';E')$, with

$$\exp(t\vec{f}(d,.)) := \sum_{n=0}^{\infty}(t^n/n!)\vec{f}(d,.)^n$$

we also have:

Corollary. With $\vec{f}$ and T' as in the corollary to Theorem 4.1 and ϕ in $F^{\infty}_{\theta',(1/\rho)}(E)$, the solution $\phi_{(.)}$ of the equation in Theorem 5.1 with $\tau = 0$ is expressible as $\phi_t = \exp(t\vec{f}(d,.))\phi$,hence $\phi_{(.)}$ is in $F^1_{T'}(K;F^{\infty}_{\theta',(1/\sigma)}(E))$, with $\{(\partial/\partial t)^k\phi_t\}_{t=0} = \vec{f}(d,.)^k\phi$ and

$$||\phi_{(.)}||_{T',1} \leq ||\phi||_{\theta',(1/\rho),\infty}(\sigma - \rho)/(\sigma - \rho\, ||\vec{f}||_{\theta,\sigma,1}eT') .$$

Theorem 5.1 is dual to Theorem 4.1 in the following sense:

Proposition 5.1. Let $H(\tau,t)h := h_\tau$ be the solution of the equation in Theorem 4.1 taking the value h when $\tau = t$. Let also $\Phi(t,\tau)\phi := \phi_t$ be the solution of the equation in Theorem 5.1 taking the value ϕ at $t = \tau$: then

$$\langle\langle h,\Phi(t,\tau)\phi\rangle\rangle = \langle\langle H(\tau,t)h,\phi\rangle\rangle .$$

Proof. By writing $H(\tau,t)h = h - \int_t^\tau \vec{f}_\lambda(.,d)H(\lambda,t)hd\lambda$ and applying Leibniz's rule (we are in a fixed Banach space) one verifies that $\tau \mapsto (\partial/\partial t)H(\tau,t)h$ satisfies the equation in Theorem 4.1 in τ, with $\{(\partial/\partial t)H(\tau,t)h\}_{\tau = t} = \vec{f}_t(.,d)h$. It follows that $(\partial/\partial t)H(\tau,t)h = H(\tau,t)\vec{f}_t(.,d)h$. Defining $H(\tau,t)'$ by

$$\langle\langle h,H(\tau,t)'\phi\rangle\rangle = \langle\langle H(\tau,t)h,\phi\rangle\rangle$$

for every h in $F^1_{\theta,\sigma}(E')$ we get

$$\langle\langle h,(\partial/\partial t)H(\tau,t)'\phi\rangle\rangle = \langle\langle(\partial/\partial t)H(\tau,t)h,\phi\rangle\rangle =$$

$$= \langle\langle H(\tau,t)\vec{f}_t(.,d)h,\phi\rangle\rangle = \langle\langle h,\vec{f}_t(.,d)H(\tau,t)'\phi\rangle\rangle$$

so $t \to H(\tau,t)'$ satisfies the equation of Theorem 5.1 with $\{H(\tau,t)'\phi\}_{t=\tau} = \phi$, that is, $H(\tau,t)'\phi = \phi_t = \Phi(t,\tau)\phi$.

Remark; Conversely, Proposition 5.1 could be used to prove Theorem 5.1 from Theorem 4.1: cf [Gelfand and Shilov 1, Chapter II, section 2].

6. Analytic dynamical systems in E'. Nonlinear dynamical systems evolving in E', governed by a vector field $\vec{f}_t$ in $F^1_{\theta,\sigma}(E';E')$, are equivalent to linear dynamical systems evolving in $F^{\infty}_{\theta',(1/\sigma}(E)$, governed by $\vec{f}_t(d,.)$, and to linear systems evolving in $F^1_{\theta,\rho}(E')$, governed by $\vec{f}$ (.,d) backwards in time, in the sense below.

Theorem 6.1. Given ξ in E' and $\vec{f}_{(.)}$ in $C^m[D_T(\tau);F^1_{\theta,\sigma}(E';E')]$, with $||\vec{f}_t||_{\theta,\sigma,1} < C$ for $|t - \tau| < T$, let ξ_t in E' be the state at time t (but t may be complex here) of a dynamical system governed by the state equation

$$(\partial/\partial t)\xi_t = \vec{f}_t(\xi_t)$$

for $|t - \tau| < T$, such that $\xi_\tau = \xi$. Given also h in $F^1_{\theta,\sigma}(E')$, let $Y(t,\tau)(\xi) := h(\xi_t)$ be a measurement of the state of the system. With the notation of Proposition 5.1 we have

$$Y(t,\tau)(\xi) = h(d)\exp<.,\xi_t>>(0) = H(\tau,t)h(\xi)$$

(h(d) as in [Dwyer 2]: cf the Introduction), hence $t \mapsto Y(t,\tau)(.)$ is in $C^{m+1}[D_T(\tau);F^1_{\theta,\rho}(E')]$, with

$$Y(t,\tau)(\xi) \le ||Y(t,\tau)(.)||_{\theta,\rho,1} \le ||h||_{\theta,\sigma,1}(\sigma - \rho)/(\sigma - \rho - eC|t - \tau|)$$

if $|t - \tau| < T \le (\sigma - \rho)/eC$ and $||\xi|| \le \rho < \sigma$.

The function space linearizations of the nonlinear problem come from the following lemma:

Lemma 6.1. Given $\vec{f}_t$ and T as before, as well as ξ in E' with $||\xi|| \le \rho$ and $t \mapsto \xi_t$ in $C^{m+1}[D_T(\tau);E']$, then $\exp<.,\xi_t>$ coincides with the solution ξ_t of the (linear) equation of Theorem 5.1, $\xi_\tau = \exp<.,\xi>$, if and only if ξ_t satisfies the (nonlinear) equation of Theorem 6.1 with $\xi_\tau = \xi$.

Proof. If $(\partial/\partial t)\xi_t = \vec{f}_t(\xi_t)$ then $||\xi_t|| \le \sigma$ for $|t-\tau|<T$ (since then ξ_t is in the domain of $\vec{f}_t$, hence $\exp<.,\xi_t>$ is in $F^\infty_{\theta',(1/\sigma)}(E)$ (and $\exp<.,\xi>$ in $F_{\theta',(1/\rho)}(E)$ by hypothesis on ξ). Moreover, $(\partial/\partial t)\exp<x,\xi_t> = [(\partial/\partial t)<x,\xi_t>]\exp<x,\xi_t> = \vec{f}_{tx}(\xi_t)\exp<x,\xi_t> =: <<\vec{f}_{tx},\exp<.,\xi_t>>>\exp<x,\xi_t> = <<\vec{f}_{tx},T_{-x}\exp<.,\xi_t>>> =: \vec{f}_t(d,x)\exp>x,\xi_t>$ (and $\exp<.,\xi_t> = \exp<.,\xi>$ if $\xi_\tau = \xi$), hence $\exp<.,\xi_t> = \phi_t$ by the uniqueness of ϕ_t. Conversely, if $\exp<.,\xi_t>$ satisfies the equation of Theorem 5.1 then $\exp<.,\xi_t>$ is in $F^\infty_{\theta',(1/\sigma)}(E)$, hence $||\xi_t|| \le \sigma$ for $|t-\tau|<T$, $\exp<.,\xi_\tau> = \exp<.,\xi>$, hence $\xi_\tau = \xi$, and $(\partial/\partial t)\xi_t = [(\partial/\partial t)\xi_t]\exp<.,\xi_t>\exp<.,-\xi_t> = [(\partial/\partial t)\exp<.,\xi_t>]\exp<.,-\xi_t> = [\vec{f}_t(d,.)\exp<.,\xi_t>]\exp<.,-\xi_t> = \ldots = [\vec{f}_t(\xi_t)\exp<.,\xi_t>]\exp<.,-\xi_t> = \vec{f}_t(\xi_t)$.

Proof of Theorem 6.1. By hypothesis on ξ_t, from Lemma 6.1 and with the notation of Proposition 5.1 we get $\exp<.,\xi_t> = \Phi(t,\tau)\exp<.,\xi>$. From Proposition 5.1 we then have $Y(t,\tau)(\xi) := h(\xi_t) =: <<h,\exp<.,\xi_t>>> = <<h,\Phi(t,\tau)\exp<.,\xi>>> = <<H(\tau,t)h,\exp<.,\xi>>> := H(\tau,t)h(\xi)$ (and $<<h,\phi>> = H(d)\phi(0)$), then use Theorem 4.1.

Corollary. Let $\vec{f}_t = f$ in $F^1_{\theta,\sigma}(E';E')$ for all t in $D_T(0)$, with $T \leq \sigma/e||\vec{f}||_{\theta,\sigma,1}$, and let ξ_t be the state of an autonomous system evolving in E' according to the state equation $(\partial/\partial t)\xi_t = \vec{f}(\xi_t)$, such that $\xi_0 = 0$. Given h in $F^1_{\theta,\sigma}(E')$ let $y(t) := h(\xi_t)$ be a measurement of the zero input response of the system: then $y(t) = \exp(tf(.,d))h(0)$, hence $y(.)$ is in $F^1_{T'}(K)$ for any $T' < T$, with $\{(\partial/\partial t)^n y(t)\}_{t=0} = \vec{f}(.,d)^n h(0)$ and

$$||y(.)||_{T'} \leq ||h||_{\theta,\sigma,1}\sigma/(\sigma - ||\vec{f}||_{\theta,\sigma,1}T').$$

Proof. From Theorem 6.1 we have $y(t) := h(\xi_t) =: Y(t,0)(0) = H(0,t)h(0)$, and from the corollary to Theorem 4.1 we have $H(0,t)h(0) = \exp(t\vec{f}(.,d)h(0)$. Moreover, if $||\xi|| \leq \rho$ then $||\exp<.,\xi>||_{\theta,(1/\rho),\infty} = 1$, hence

$$|H(0,t)h(\xi)| = |<<H(0,t)h,\exp<.,\xi> >>| \leq$$

$$\leq ||H(0,t)h||_{\theta,\rho,1} \leq ||h||'_{\theta,\sigma,1}(\sigma - \rho)/(\sigma - \rho - e||\vec{f}||_{\theta,\sigma,1}|t|)$$

for $|t| < (\sigma - \rho)/e||f||_{\theta,\sigma,1}$. If $\xi = 0$ then we may let ρ tend to 0, getting the desired estimate.

References

N. Ahmed, [1]: Strong and weak synthesis of nonlinear systems with constraints on the system space G_λ, Information and Control 23 (1973), 71-85.

J. Baillieul, [1]: Multilinear optimal control, Proceedings of the 1976 Ames (NASA) Conference on Geometric Control, R. Hermann and C. Martin, Editors, Math Sci Press, 53 Jordan Road, Brookline, MA 02138, U.S.A.

R. Brockett, [1]: Nonlinear Systems and differential geometry, Proc. IEEE 64 (1976), 61-72.

[2]: Volterra series and geometric control theory, Automatica 12 (1976), 167-176.

T. Dwyer, [1]: Dualite des espaces de fonctions entieres en dimension infinie, Ann. Inst. Fourier (Grenoble), 26 (1976), 151-195.

[2]: Differential operators of infinite order in locally convex spaces, Part I, Rendiconti di Matematica, 10 (1977), 1-30.

[3]: Differential equations of infinite order in vector-valued holomorphic Fock spaces, to appear in Infinite-Dimensional Holomorphy and Aplications, M. Matos, Editor, North-Holland Mathematics Studies, North-Holland Publishing Co., Amsterdam.

[4]: Fourier-Borel duality and bilinear realizations of control systems, Proceedings of the 1976 Ames (NASA) Conference on Gepmetric Control, R. Hermann and C Martin, Editors, Math Sci Press, 53 Jordan Road, Brookline, MA 02138.

I. Gelfand and G. Shilov, [1]: Generalized Functions, vol. 3, Theory of Differential Equations, Academic Press, New York, 1967.

L. Nachbin, [1]: Topology on Spaces of Holomorphic Mappings, Ergebnisse fur die Mathematik und ihrer Grenzgebiete, Band 47, Springer-Verlag, Berlin 1969.

F. Trèves, [1]: Ovcyannikov Theorem and Hyperdifferential Operators, Notas de Matemática N° 46, Instituto de Matemática Pura e Aplicada, Rio de Janeiro, 1968.

[2]: Basic Linear Partial Differential Equations, Academic Press, New York, 1975.

W. Walter, [1]: Approximation für das Cauchy-Problem bei parabolischen Differentialgleichungen mit der Linienmethode, Abstract Spaces and Approx., Proc. Conf. Math. Res. Inst. Oberwolfach 1968, 375-385 (1969).

Northern Illinois Univ.,
Dekalb, Illinois 60115,
U.S.A.

ON THE RADON-NIKODYM-PROPERTY AND MARTINGALE CONVERGENCE*

G. A. Edgar
Department of Mathematics
The Ohio State University
Columbus, Ohio 43210 U.S.A.

It is well-known that there is an intimate connection between the Radon-Nikodym property and martingale convergence in a Banach space. This connection can be "localized" to a closed bounded convex subset of a Banach space. In this paper we are interested primarily in this connection for a bounded convex set which is not closed.

If C is a bounded convex set in a locally convex space, C is said to have the martingale convergence property iff every martingale with values in C converges in measure. Since C is not assumed to be metrizable, it is appropriate to use martingales indexed by an arbitrary directed set, and not restrict attention to sequential martingales. Similarly, C is said to have the Radon-Nikodym property iff every vector-valued measure defined on a probability space with average range in C has a derivative which has sufficiently strong measurability properties. The one-dimensional example of an open interval shows that the two properties are no longer equivalent. Theorem 2.4 describes the connection between the two notions.

This paper is also concerned with an ordering on the tight probability measures on a bounded convex set C. The ordering $\prec$, which has been called "comparison of experiments", "the Choquet ordering", "the dilation ordering", and many others, can be described in many equivalent ways; they are given in Theorem 2.2. For example, $\mu \prec \nu$ means $\int f d\mu \leq \int f d\nu$ for all bounded continuous convex functions f on C. Other descriptions of the ordering involve dilations and conditional expectations. Earlier versions of this theorem have been attributed to: Hardy, Littlewood and Polya, Blackwell, Stein, Sherman, Cartier and Strassen.

One other result proved here deserves mention (Corollary 2.7). If C is a separable closed bounded convex subset of a Banach space, and if each point of C admits a unique representing measure on the extreme points of C, then C has the Radon-Nikodym property.

---

* Supported in part by National Science Foundation grant MCS77-04049.

The attentive reader will notice that the assumption of convexity is hardly ever used in an essential way, so that most of what appears below has a reformulation for nonconvex sets C. I have not included such reformulations here.

The paper has two sections. The first is preliminary; the results there are either substantially known or straightforward, so most proofs are omitted. The second section contains the main results of the paper, including those mentioned above; proofs are given in this section.

1.

If T is a completely regular Hausdorff space, we will write $\mathcal{C}_b(T)$ for the Banach space of all bounded continuous real-valued functions on T. We will be interested in several subsets of the dual $\mathcal{C}_b(T)^*$ in its weak* topology. First,

$$\mathcal{P}_f(T) = \{\mu \in \mathcal{C}_b(T)^* : \langle \mu, 1 \rangle = 1 \, , \; \mu \geq 0\} \; ;$$

this can be identified with the set of finitely-additive regular probability measures on the algebra generated by the zero sets [18, p. 165]; the identification (and similar ones below) will be made whenever convenient. Note that $\mathcal{P}_f(T)$ is compact. Next,

$$\mathcal{P}_\sigma(T) = \{\mu \in \mathcal{P}_f(T) : \text{if} \;\; f_n \in \mathcal{C}_b(T) \; (n = 1, 2, \ldots), \; f_n \downarrow 0 \, , \text{then} \; \langle \mu, f_n \rangle \to 0\} \; ;$$

these measures extend uniquely to the Baire sets of T. Also,

$$\mathcal{P}_\tau(T) = \{\mu \in \mathcal{P}_f(T) : \text{if} \;\; f_\alpha \in \mathcal{C}_b(T) \text{ is a net, } f_\alpha \downarrow 0 \, , \text{then} \;\; \langle \mu, f_\alpha \rangle \to 0\} \; ;$$

these measures extend uniquely to (closed-) regular measures on the Borel sets of T. These are called τ-<u>smooth</u> measures. Next,

$$\mathcal{P}_t(T) = \{\mu \in \mathcal{P}_\tau(T) : \text{for each } \; \epsilon > 0 \, , \text{ there is a compact set } K \subseteq T \;\; \text{such that} \;\; \mu(K) \geq 1 - \epsilon\} \; ;$$

these are called <u>tight</u> measures (on the Baire sets) or <u>Radon</u> measures (on the Borel sets). Finally,

$$\mathcal{P}_s(T) = \{\sum_{i=1}^{\infty} t_i \epsilon_{x_i} : t_i \geq 0 \, , \; \Sigma \, t_i = 1 \, , \; x_i \in T\} \; ,$$

$$\mathcal{P}_d(T) = \{\sum_{i=1}^{n} t_i \epsilon_{x_i} : n \in \mathbb{N} \, , \; t_i \geq 0 \, , \; \Sigma \, t_i = 1 \, , \; x_i \in T\} \; .$$

Note $\mathcal{P}_f \supseteq \mathcal{P}_\sigma \supseteq \mathcal{P}_\tau \supseteq \mathcal{P}_t \supseteq \mathcal{P}_s \supseteq \mathcal{P}_d$, with $\mathcal{P}_f = \mathcal{P}_t$ if T is compact.

Let E be a locally convex (Hausdorff) topological vector space, and C a subset of E. If $\mu \in \mathcal{P}_f(C)$ and $x \in E$, we say that x is the <u>resultant</u> of μ, and write $x = r(\mu)$, iff for every $f \in E^*$, we have $\langle \mu, f \rangle = f(x)$. The set C will be called d-convex [resp. s-, t-, τ-, σ-, f-convex] iff for every $\mu \in \mathcal{P}_d(C)$ [resp. $\mathcal{P}_s(C)$, etc.], there exists $r(\mu) \in C$. Note that d-convex is the same as convex and that f-convex is the same as compact and convex. We will say that C satisfies <u>condition (EC)</u> iff the closed convex hull of a compact subset of C is a compact subset of C, i.e. if $K \subseteq C$ is compact, then there is a compact convex set K_1 with $K \subseteq K_1 \subseteq C$.

The following is from [8].

1.1 PROPOSITION. (a) <u>The set</u> C <u>satisfies condition (EC) if and only if, for every measure</u> $\mu \in \mathcal{P}_f(C)$ <u>with compact support, the resultant</u> $r(\mu)$ <u>exists in</u> C.

(b) C <u>is</u> t-<u>convex if and only if</u> C <u>is</u> s-<u>convex and satisfies condition (EC)</u>.

It is easy to show that $\mathcal{P}_x(T)$ is x-convex, where $x = d, s, \sigma, f$. It is not hard to show that $\mathcal{P}_\tau(T)$ is τ-convex. Indeed, suppose $\gamma \in \mathcal{P}_\tau(\mathcal{P}_\tau(T))$. Then $\mu = r(\gamma)$ exists in $\mathcal{P}_f(T)$. If f_α is a net in $\mathcal{C}_b(T)$ with $f_\alpha \downarrow 0$, then for all $\lambda \in \mathcal{P}_\tau(T)$, we have $\langle \lambda, f_\alpha \rangle \downarrow 0$. But for each α, the function $\lambda \mapsto \langle \lambda, f_\alpha \rangle$ is in $\mathcal{C}_b(\mathcal{P}_\tau(T))$, so $\langle \mu, f_\alpha \rangle = \int \langle \lambda, f_\alpha \rangle \, d\gamma(\lambda) \to 0$ since γ is τ-smooth. Thus μ is τ-smooth.

An example of D. H. Fremlin shows that $\mathcal{P}_t(T)$ need not satisfy condition (EC) and therefore need not be t-convex. Clearly, if $\mathcal{P}_\tau(T) = \mathcal{P}_t(T)$ (such spaces T are called, variously, "universally measurable" or "semi-Radonian" [10, Theorem 2, p. 133]), then $\mathcal{P}_t(T)$ is t-convex.

Let C be a subset of a locally convex space E, and let $(\Omega, \mathfrak{F}, P)$ be a probability space. If $\varphi: \Omega \to C$ is Borel measurable, we define a Borel measure $\varphi(P)$ on C by $\varphi(P)(B) = P(\varphi^{-1}(B))$. We will write $L^0(\Omega, \mathfrak{F}, P; C)$ for the set of all Borel measurable functions $\varphi: \Omega \to C$ such that $\varphi(P) \in \mathcal{P}_t(C)$, i.e. for every Borel set $B \subseteq C$, and every $\epsilon > 0$, there is a compact set $K \subseteq B$ with $P(\varphi^{-1}(B) \setminus \varphi^{-1}(K)) < \epsilon$. For $\varphi \in L^0(\Omega; C)$, we will write $x = \int_A \varphi \, dP$ iff $f(x) = \int_A f(\varphi(\omega)) dP(\omega)$ for all $f \in E^*$; if such an element x exists for each $A \in \mathfrak{F}$, we will say that φ is Pettis integrable. (Elements φ, ψ of L^0 should be identified iff they are <u>weakly equivalent</u>, i.e. $f \circ \varphi = f \circ \psi$ a.e. for all $f \in E^*$, the exceptional set may depend on f.)

Let E be a locally convex space, let $(\Omega, \mathfrak{F}, P)$ be a probability space, and let $m: \mathfrak{F} \to E$ be a vector-valued measure. The P-average range of m is $\{m(A)/P(A): A \in \mathfrak{F}, P(A) > 0\}$. We say m is differentiable with respect to P iff there exists $\varphi \in L^0(\Omega;E)$ such that $m(A) = \int_A \varphi\, dP$ for all $A \in \mathfrak{F}$; in that case we write $\varphi = dm/dP$. A bounded subset C of E is said to have the Radon-Nikodym property iff for any probability space $(\Omega, \mathfrak{F}, P)$ and any measure $m: \mathfrak{F} \to E$ with $m \ll P$ and average range contained in C, there exists $dm/dP \in L^0(\Omega, \mathfrak{F}, P;C)$.

Here is the Radon-Nikodym theorem which will be used below. The basic form goes back to Grothendieck; the version given here can be found in [13, Theorem 4.9], except for the assertion that $dm/dP \in L^0$.

1.2 THEOREM. Let C be a subset of a locally convex space E, let $(\Omega, \mathfrak{F}, P)$ be a complete probability space, and let $m: \mathfrak{F} \to E$ be a vector-valued measure $\ll P$. Suppose that m almost has P-average range relatively compact in C. Then there exists $dm/dP \in L^0(\Omega, \mathfrak{F}, P;C)$.

The following corollary has been proved independently by several mathematicians (see for example [4, Theorem 3.1], [9, Théorème 4.2]).

1.3 COROLLARY. Let T be a completely regular space. Then $\mathcal{P}_t(T)$ has the Radon-Nikodym property.

Let E be a locally convex space, $(\Omega, \mathfrak{F}, P)$ a probability space, $\varphi \in L^0(\Omega, \mathfrak{F}, P;E)$. Let $\mathcal{B}$ be a sub-σ-algebra of $\mathfrak{F}$. A conditional expectation of φ given $\mathcal{B}$ is a function $\psi \in L^0(\Omega, \mathcal{B}, P;E)$ such that $f \circ \psi = E[f \circ \varphi | \mathcal{B}]$ for all $f \in E^*$; we write $\psi = E[\varphi|\mathcal{B}]$. If $\varphi \in L^0(\Omega;C)$, where C is a bounded t-convex subset of E, then $\psi = E[\varphi|\mathcal{B}]$ if and only if $\int_A \psi\, dP = \int_A \varphi\, dP$ for all $A \in \mathcal{B}$. Thus, if C also has the Radon-Nikodym property, then $E[\varphi|\mathcal{B}]$ necessarily exists.

Let C be a bounded convex subset of a locally convex space E. A martingale in C consists of: a probability space $(\Omega, \mathfrak{F}, P)$; a directed set J; a family $(\mathfrak{F}_\alpha)_{\alpha \in J}$ of sub-σ-algebras of $\mathfrak{F}$ indexed by J such that $\mathfrak{F}_\alpha \subseteq \mathfrak{F}_\beta$ if $\alpha \leq \beta$; and a family $(\varphi_\alpha)_{\alpha \in J}$ where $\varphi_\alpha \in L^0(\Omega, \mathfrak{F}_\alpha, P;C)$ and $\varphi_\alpha = E[\varphi_\beta | \mathfrak{F}_\alpha]$ if $\alpha \leq \beta$. Let $(\varphi_\alpha)_{\alpha \in J}$ be a martingale in C, and let $\varphi \in L^0(\Omega, \mathfrak{F}, P;C)$. We say that φ closes (φ_α) iff $\varphi_\alpha = E[\varphi|\mathfrak{F}_\alpha]$ for all $\alpha \in J$. We say that φ_α converges in measure to φ iff, for every neighborhood U of 0 in E,

$$\lim_{\alpha \in J} P\{\omega: \varphi_\alpha(\omega) - \varphi(\omega) \in U\} = 1 .$$

We say that φ_α converges in mean to φ iff, for every continuous seminorm q on E,

$$\lim_{\alpha \in J} \int q(\varphi_\alpha(\omega) - \varphi(\omega))\,dP(\omega) = 0 .$$

1.4 PROPOSITION. Let (φ_α) be a martingale in C, and let $\varphi \in L^0(\Omega, \mathfrak{F}, P; C)$. The following are equivalent.

(a) φ_α converges in measure to φ;

(b) φ_α converges in mean to φ;

(c) φ closes (φ_α) and φ is measurable with respect to the σ-algebra generated by $\bigcup_{\alpha \in J} \mathfrak{F}_\alpha$.

1.5 PROPOSITION. Let (φ_α) be a martingale in C. Then (φ_α) converges in measure if and only if $\varphi_\alpha(P)$ converges in $\mathcal{P}_t(C)$.

The bounded convex set C is said to have the J-martingale convergence property iff every martingale in C indexed by J converges in measure. (The $\mathbb{N}$-martingale convergence property will be called the sequential martingale convergence property.) Also, C is said to have the martingale convergence property iff it has the J-martingale convergence property for all directed sets J. The well-ordered martingale convergence property and the totally-ordered martingale convergence property are defined in a similar fashion.

2.

Let E be a locally convex space, and let C be a bounded convex subset of E. A partial order can be defined on $\mathcal{P}_t(C)$ as follows. If $\mu, \nu \in \mathcal{P}_t(C)$, define $\mu \prec \nu$ iff $\langle \mu, f\rangle \leq \langle \nu, f\rangle$ for all bounded continuous convex functions f on C. This relation is clearly reflexive and transitive; the antisymmetry of the relation follows from the following result, which essentially goes back to LeCam (see [14, p. 216], [11, Lemma 2.1], [12, Lemma 3.1]).

2.1 PROPOSITION. Let $(T, \mathcal{T})$ be a completely regular space. Suppose $F \subseteq \mathcal{C}_b(T)$ is a class of functions which separates points of T and if $f, g \in F$, then the pointwise maximum $f \vee g \in F$. Let $\mathcal{T}'$ be the topology on T generated by F. If $\mu_\alpha, \mu \in \mathcal{P}_t(T, \mathcal{T}')$ and

$$\lim_\alpha \int_T f d\mu_\alpha = \int_T f d\mu \quad \text{for all} \quad f \in F,$$

then $\mu_\alpha \to \mu$ in the weak* topology of $\mathcal{P}_t(T, \mathcal{T}')$. In particular, if $\mu, \nu \in \mathcal{P}_t(T, \mathcal{T})$ and $\int f d\mu = \int f d\nu$ for all $f \in F$, then $\mu = \nu$.

Let $\mu \in \mathcal{P}_t(C)$. A μ-dilation is a function

$$T \in L^0(C, \mathcal{B}^\mu(C), \mu; \mathcal{P}_t(C))$$

such that for every $h \in E^*$, we have $\langle T(x), h\rangle = h(x)$ for μ-almost every x. ($\mathcal{B}(C)$) denotes the Borel sets on C; $\mathcal{B}^\mu(C)$ the completion with respect to μ;

$\mathcal{P}_t(C)$ is understood to have its weak* topology.) If C is t-convex, then $r(T(x))$ exists for every x, so in that case the condition is the same as the assertion that $r \circ T$ is weakly equivalent to the identity on C. If $\nu \in \mathcal{P}_t(C)$, we write $\nu = T(\mu)$, and say that μ dilates to ν, iff $\langle \nu, f \rangle = \int \langle T(x), f \rangle \, d\mu(x)$ for all $f \in \mathcal{C}_b(C)$; that is, $\nu = \int T \, d\mu$ in $\mathcal{P}_t(C)$.

The following theorem shows that the ordering $\prec$ can be characterized in terms of dilations and in terms of conditional expectations. It goes back (in the one-dimensional case) to Hardy, Littlewood and Polya. The proof here is closest to that of V. Strassen [17].

2.2. THEOREM. Let C be bounded and convex, and let $\mu, \nu \in \mathcal{P}_t(C)$. The following are equivalent:

(a) $\mu \prec \nu$; i.e. $\langle \mu, f \rangle \leq \langle \nu, f \rangle$ for all bounded continuous convex functions f on C;

(b) μ dilates to ν; i.e. there is a μ-dilation $T: C \to \mathcal{P}_t(C)$ such that $T(\mu) = \nu$;

(c) There exist a probability space $(\Omega, \mathfrak{F}, P)$, a σ-algebra $\mathcal{G} \subseteq \mathfrak{F}$, and functions $\varphi, \psi \in L^0(\Omega, \mathfrak{F}, P; C)$ with $\varphi(P) = \mu$, $\psi(P) = \nu$, and $\varphi = E[\psi | \mathcal{G}]$.

Proof. (a) $\Rightarrow$ (b). If $f \in \mathcal{C}_b(C)$ define its upper envelope $f\hat{}: C \to \mathbb{R}$ by

$$f\hat{}(x) = \inf\{h(x) : h \text{ is bounded continuous and affine on } C \text{ and } h \geq f\}.$$

Note that for fixed f, the map $x \mapsto f\hat{}(x)$ is upper semicontinuous, hence Borel measurable. Also, for fixed x, the map $f \mapsto f\hat{}(x)$ is continuous for the uniform norm on $\mathcal{C}_c(C)$. We can also write

$$f\hat{}(x) = \inf\{h(x) : h \text{ is bounded continuous and concave on } C \text{ and } h \geq f\}.$$

Let S be the vector space of all Borel-measurable simple functions θ: $X \to \mathcal{C}_b(C)$. Define $p: S \to \mathbb{R}$ by

$$p(\theta) = \int \theta(x)\hat{}(x) \, d\mu(x). \tag{1}$$

This integral exists since the integrand is bounded and Borel measurable. It is easily checked that p is a sublinear functional on S. For $f \in \mathcal{C}_b(C)$ and $A \in \mathcal{B}(C)$, define $\chi_A \otimes f \in S$ by

$$(\chi_A \otimes f)(x) = \begin{cases} f & \text{if } x \in A \\ 0 & \text{if } x \notin A. \end{cases}$$

Thus $f \mapsto \chi_C \otimes f$ is an embedding of $\mathcal{C}_b(C)$ in S. Define a linear functional ℓ on the range of this embedding by

$$\ell(\chi_C \otimes f) = \langle \nu, f \rangle . \tag{2}$$

We claim that $\ell(\chi_C \otimes f) \leq p(\chi_C \otimes f)$. Let h be concave bounded and continuous on C and $h \geq f$. Then $\int f \, d\nu \leq \int h \, d\nu \leq \int h \, d\mu$ (by the assumption that $\mu \prec \nu$), so by the tightness of μ, $\int f \, d\nu \leq \int f\hat{}\, d\mu$; i.e. $\ell(\chi_C \otimes f) \leq p(\chi_C \otimes f)$.

By the Hahn-Banach theorem, the linear functional ℓ can be extended to all of S with $\ell(\theta) \leq p(\theta)$ for all $\theta \in S$. Define $m: \mathcal{B}(C) \to \mathcal{C}_b(C)^*$ by $\langle m(A), f \rangle = \ell(\chi_A \otimes f)$. Now if h is affine, bounded and continuous on C, then $h\hat{} = h$, so $\langle m(A), h \rangle = \ell(\chi_A \otimes h) \leq p(\chi_A \otimes h) = \int_A h\hat{}\, d\mu = \int_A h d\mu$; similarly $(-h)\hat{} = -h$, so $\langle m(A), -h \rangle \leq \int_A (-h) d\mu$; therefore

$$\langle m(A), h \rangle = \int_A h \, d\mu . \tag{3}$$

In particular $\langle m(A), 1 \rangle = \mu(A)$. Again, if $f \in \mathcal{C}_b(C)$ and $f \geq 0$, then $(-f)\hat{} \leq 0$, so $\langle m(A), f \rangle = \ell(\chi_A \otimes (-f)) \leq p(\chi_A \otimes (-f)) = \int_A (-f)\hat{}\, d\mu \leq 0$; thus $\langle m(A), f \rangle \geq 0$. Now if $f \geq 0$, then $\langle m(A), f \rangle = \langle m(C), f \rangle - \langle m(C \backslash A), f \rangle \leq \langle m(C), f \rangle = \ell(\chi_C \otimes f) = \langle \nu, f \rangle$, so $0 \leq m(A) \leq \nu$. This shows that $m(A)$ is tight, so $m(A)/\mu(A) \in \mathcal{P}_t(C)$. Thus the vector-valued measure $m: \mathcal{B}(C) \to \mathcal{C}_b(C)^*$ has average range in $\mathcal{P}_t(C)$.

By Corollary 1.3, there is $T: C \to \mathcal{P}_t(C)$, $T \in L^0(C, \mathcal{B}^\mu(C), \mu; \mathcal{P}_t(C))$, such that $\int_A T \, d\mu = m(A)$ for all $A \in \mathcal{B}(C)$. Now $\langle \nu, f \rangle = \ell(\chi_C \otimes f) = \langle m(C), f \rangle = \int \langle T(x), f \rangle \, d\mu(x)$ for all $f \in \mathcal{C}_b(C)$, i.e. $\nu = T(\mu)$. Finally, if $h \in E^*$, then for $A \in \mathcal{B}(C)$, using (3) yields $\int_A \langle T(x), h \rangle \, d\mu(x) = \langle m(A), h \rangle = \int_A h \, d\mu$, so $\langle T(x), h \rangle = h(x)$ μ-a.e. Thus T is a μ-dilation.

(b) $\Rightarrow$ (c). Suppose $\nu = T(\mu)$. Let $\Omega = C \times C$, $\mathfrak{F} = \mathcal{B}(C) \times \mathcal{B}(C)$, $\mathcal{G} = \mathcal{B}(C) \times \{C, \emptyset\}$, and define $\varphi, \psi: \Omega \to C$ by $\varphi(x,y) = x$, $\psi(x,y) = y$. Define P on $\mathfrak{F}$ by

$$P(D) = \int T(x)(D_x) \, d\mu(x) ,$$

where $D_x = \{y \in C: (x,y) \in D\}$. The integrand is $\mathcal{B}^\mu(C)$ - measurable since T is a dilation. It follows that

$$\int g \, dP = \int \int g(x,y) \, d[T(x)](y) \, d\mu(x)$$

for any bounded $\mathcal{F}$-measurable function g on Ω. Now for $A \in \mathcal{B}(C)$,

$$\varphi(P)(A) = P(\varphi^{-1}(A)) = \int T(x)((A \times C)_x) d\mu(x)$$

$$= \int_A 1 \, d\mu(x) = \mu(A) ,$$

$$\psi(P)(A) = P(\psi^{-1}(A)) = \int T(x)((C \times A)_x) \, d\mu(x)$$

$$= \int T(x)(A) \, d\mu(x) = T(\mu)(A) = \nu(A) ,$$

so $\varphi(P) = \mu$, $\psi(P) = \nu$. This shows that $\varphi, \psi \in L^0(\Omega;C)$. Also, for any $A \in \mathcal{B}(C)$ and $h \in E^*$,

$$\int_{A \times C} h \circ \psi \, dP = \int_A \int_C h(y) dT(x)(y) \, d\mu(x)$$

$$= \int_A h(x) \, d\mu(x) = \int_A \int_C h(x) dT(x)(y) \, d\mu(x)$$

$$= \int_{A \times C} h \circ \varphi \, dP ,$$

so $\varphi = E[\psi|\mathcal{S}]$.

(c) ⇒ (a). Suppose $\varphi = E[\psi|\mathcal{S}]$ and $\varphi(P) = \mu$, $\psi(P) = \nu$. Let f be bounded, continuous and convex on C. They by Jensen's inequality, $\langle\mu,f\rangle = \int f \circ E[\psi|\mathcal{S}] dP \leq \int E[f \circ \psi|\mathcal{S}] dP = \int f \circ \psi \, dP = \langle\nu,f\rangle$. □

Here is a lemma which is probably known, but I am unable to provide a reference. The proof is, apparently, not merely are application of Zorn's Lemma, but requires well-ordering as well.

2.3. LEMMA. Let $(P,\leq)$ be a partially ordered set. Suppose every chain in P has at least upper bound. Then any subset of P which is directed has at least upper bound.

Proof. Let $D \subseteq P$ be a directed subset of P. Let $\mathcal{M}$ be the collection of all $A \subseteq P$ which satisfy

(a) $A \supseteq D$,

(b) if C is a chain included in A, then $\sup C \in A$.

Notice that $P \in \mathcal{M}$, so $\mathcal{M} \neq \emptyset$. Let $M = \cap \mathcal{M}$. Then $M \in \mathcal{M}$.

Next, let $\mathcal{R}$ be the collection of all $B \subseteq P$ which satisfy

(a) $D \subseteq B \subseteq M$,

(b) B is directed.

Notice that $D \in \mathcal{R}$, so $\mathcal{R} \neq \emptyset$. Apply Zorn's Lemma to $\mathcal{R}$; let $R \in \mathcal{R}$ be maximal. We claim that $R \in \mathcal{M}$. Trivially $R \supseteq D$. Suppose (for purposes of contradiction) that not every chain in R has its sup in R. Then there is a well-ordered chain in R with sup not in R. Let ξ be the least ordinal of such a chain. Let $R' = \{\sup C\colon C \text{ is a chain of order type } \xi \text{ included in } R\} \cup R$. We claim that $R' \in \mathfrak{R}$. Clearly $R' \supseteq R \supseteq D$ and $R' \subseteq M$. To prove that R' is directed, let $x,y \in R'$, say $x = \sup\{x_\gamma\colon \gamma < \xi\}$, $y = \sup\{y_\gamma\colon \gamma < \xi\}$, with $x_\gamma, y_\gamma \in R$. (If $x \in R$, let $x_\gamma = x$ for all $\gamma < \xi$.) Define $z_\gamma \in R$ inductively so that

$$z_{\gamma+1} \geq x_\gamma\,,\ z_{\gamma+1} \geq y_\gamma\,,\ z_{\gamma+1} \geq z_\gamma \tag{1}$$

and if $\gamma < \xi$ is a limit ordinal,

$$z_\gamma = \sup\{z_\beta\colon \beta < \gamma\}\ . \tag{2}$$

Now (1) is possible since R is directed and (2) is possible by the minimality of ξ. Then $z = \sup\{z_\gamma\colon \gamma < \xi\} \in R'$ and $z \geq x$, $z \geq y$. Thus R' is directed, so $R' \in \mathfrak{R}$. By the maximality of R, we have $R' = R$, so in fact every chain in R has its sup in R. Thus $R \in \mathfrak{M}$. Then we have $R \supseteq M$, so $R = M$. This chows that M is directed.

Next, we claim that every subset S of D has an upper bound in M. To prove this, well-order $S = \{x_\gamma\colon \gamma < \alpha\}$, and proceed by induction on α. For $\alpha = 1$, $S = \{x_0\} \subseteq M$. If $\alpha = \beta + 1$, let $y \in M$ be an upper bound for $\{x_\gamma\colon \gamma < \beta\}$, which exists by the induction hypothesis; since M is directed, there is an upper bound for $\{y, x_\beta\}$ in M. If α is a limit ordinal, construct inductively $y_\gamma \geq x_\gamma$, $y_\gamma \in M$, y_γ increasing. Since $M \in \mathfrak{M}$, $\sup\{y_\gamma\colon \gamma < \alpha\} \in M$ and is the required upper bound.

In particular, D itself has an upper bound in M, say x_0. If y is any upper bound for D in P, then $\{x \in P\colon x \leq y\} \in \mathfrak{M}$, so $M \subseteq \{x \in P\colon x \leq y\}$, and hence $x_0 \leq y$. Therefore x_0 is the least upper bound of D, q.e.f. $\square$

2.4. THEOREM. <u>Let C be a bounded convex set in a locally convex space. Consider the following conditions:</u>

(a) C <u>has both the Radon-Nikodym property and the sequential martingale convergence property.</u>

(b) C <u>has the well-ordered martingale convergence property.</u>

(c) C <u>has the martingale convergence property.</u>

(b') <u>Every well-ordered subset of $\mathcal{P}_t(C)$ has a least upper bound.</u>

(c') <u>Every directed subset of $\mathcal{P}_t(C)$ has a least upper bound.</u>

<u>Then</u>: (b), (c), (b'), (c') <u>are equivalent and imply</u> (a). <u>If</u> C <u>is</u> t-<u>convex, then all five conditions are equivalent.</u>

Proof. (b) $\Rightarrow$ (c). Suppose C has the well-ordered martingale convergence property. Let D be a directed set, and suppose $(\varphi_\alpha)_{\alpha \in D}$ is a martingale with values in C ; the additional data is $(\Omega, \mathfrak{F}, P)$, $(\mathfrak{F}_\alpha)_{\alpha \in D}$. There is a partially ordered set $Q \supseteq D$ in which each chain has a least upper bound, namely let Q be the set of ideals of D [1, p. 113]. As in the proof of Lemma 2.3, there is a directed set M , $D \subseteq M \subseteq Q$, such that every chain in M has sup in M , and if $D \subseteq A \subseteq M$ and every chain in A has sup in A , then $A = M$. For each $\gamma \in M$, define $\mathfrak{F}_\gamma$ to be the σ-algebra generated by $\bigcup\{\mathfrak{F}_\alpha : \alpha \in D, \alpha \leq \gamma\}$. Let $A = \{\gamma \in M$: there exists $\varphi_\gamma \in L^0(\Omega, \mathfrak{F}_\gamma, P; C)$ such that $\varphi_\alpha = E[\varphi_\gamma | \mathfrak{F}_\alpha]$ for all $\alpha \in D$ with $\alpha \leq \gamma\}$. Note that such a function φ_γ (if it exists) is unique as an element of $L^0(\Omega, \mathfrak{F}_\gamma, P; C)$. Hence, if $\gamma, \gamma' \in A$, and $\gamma \leq \gamma'$, then $\varphi_\gamma = E[\varphi_{\gamma'} | \mathfrak{F}_\gamma]$.

Clearly $D \subseteq A \subseteq M$. We claim that every chain in A has sup in A . Let B be a chain included in A ; write $\gamma_0 = \sup B$. Now B has a well-ordered cofinal subset B_0 . Then $(\varphi_\gamma)_{\gamma \in B_0}$ is a well-ordered martingale, so it converges; by Proposition 1.4, its limit φ_{γ_0} satisfies $\varphi_\alpha = E[\varphi_{\gamma_0} | \mathfrak{F}_\alpha]$ for all $\alpha \in D$, $\alpha \leq \gamma_0$, so $\gamma_0 \in A$. (By the choice of Q , if $\alpha \leq \gamma_0$, then $\alpha \leq \gamma$ for some $\gamma \in B_0$.) Thus, every chain in A has sup in A , so we have $A = M$.

Now M is directed, so M has a largest element γ^* . The martingale $(\varphi_\alpha)_{\alpha \in D}$ is closed by φ_{γ^*} , so (φ_α) converges.

(c) $\Rightarrow$ (b) is trivial.

(c) $\Rightarrow$ (a). Suppose C has the martingale convergence property. Then C trivially has the sequential martingale convergence property. Let $(\Omega, \mathfrak{F}, P)$ be a probability space and $m: \mathfrak{F} \to E$ a vector-valued measure with $m \ll P$ and average range included in C . Let D be the set of finite $\mathfrak{F}$-measurable partitions of Ω , ordered by a.e. refinement. For $\alpha = \{A_1, A_2, \ldots, A_n\} \in D$, let $\mathfrak{F}_\alpha$ be the σ-algebra generated by $\{A_1, \ldots, A_n\}$, and define $\varphi_\alpha : \Omega \to E$ by

$$\varphi_\alpha = \sum_{A \in \alpha} \frac{m(A)}{P(A)} \chi_A ,$$

where we interpret $0/0$ as an arbitrary element of C . Then $\varphi_\alpha \in L^0(\Omega, \mathfrak{F}_\alpha, P; C)$, and if $\alpha \leq \beta$, we have $\varphi_\alpha = E[\varphi_\beta | \mathfrak{F}_\alpha]$. Thus (φ_α) is a martingale with values in C . Let $\varphi \in L^0(\Omega, \mathfrak{F}, P; C)$ be the limit of (φ_α) . Thus $\varphi_\alpha = E[\varphi | \mathfrak{F}_\alpha]$ for all $\alpha \in D$ by Proposition 1.4. Let $A \in \mathfrak{F}$. Then $\alpha = \{A, \Omega \setminus A\} \in D$, and we have

$$m(A) = \frac{m(A)}{P(A)} \cdot P(A) = \int_A \varphi_\alpha \, dP = \int_A \varphi \, dP ,$$

so $dm/dP = \varphi$. Thus C has the Radon-Nikodym property.

(a) ⇒ (b). Let C be t-convex. Suppose C has the Radon-Nikodym property and the sequential martingale convergence property. Let $(\varphi_\alpha)_{\alpha<\xi}$ be a martingale indexed by the ordinals less than the ordinal ξ . If $\xi = \beta + 1$, then $(\varphi_\alpha)_{\alpha<\xi}$ is closed by φ_β . If ξ has countable cofinality, then convergence follows from the sequential martingale convergence property. Suppose that ξ has uncountable cofinality, that is, any countable set of ordinals $< \xi$ has an upper bound $< \xi$. Let $\mathfrak{F}_\infty$ denote the σ-algebra generated by $\bigcup\{\mathfrak{F}_\alpha : \alpha < \xi\}$. Since ξ has uncountable cofinality, $\mathfrak{F}_\infty = \bigcup\{\mathfrak{F}_\alpha : \alpha < \xi\}$. For $A \in \mathfrak{F}_\infty$, say $A \in \mathfrak{F}_\alpha$, let $m(A) = \int_A \varphi_\alpha \, dP$. This integral exists since C is t-convex. The average range of m is included in C , so by the Radon-Nikodym property, there exists $\varphi = dm/dP$. Now if $A \in \mathfrak{F}_\alpha$, $\int_A \varphi \, dP = m(A) = \int_A \varphi_\alpha dP$, so $\varphi_\alpha = E[\varphi|\mathfrak{F}_\alpha]$. Thus $(\varphi_\alpha)_{\alpha<\xi}$ converges to φ .

(b') ⇒ (c') follows from Lemma 2.3; (b') ⇒ (b) follows from Proposition 1.5.

(c') ⇒ (b') is trivial.

(b) ⇒ (b'). Suppose C has the well-ordered martingale convergence property. Let $(\mu_\gamma)_{\gamma<\xi}$ be a well-ordered subset of $\mathcal{P}_t(C)$. Fir $\alpha \leq \xi$, let Ω_α be a product of copies of C indexed by the ordinals $< \alpha$; equip Ω_α with the product σ-algebra; for $\gamma < \alpha \leq \xi$, define $\varphi_{\alpha,\gamma} : \Omega_\alpha \to C$ by $\varphi_{\alpha,\gamma}(\omega) = \omega(\gamma)$; and let $\mathfrak{F}_{\alpha,\gamma}$ be the σ-algebra on Ω_α generated by $(\varphi_{\alpha,\beta})_{\beta\leq\gamma}$. We will define inductively a measure P_α on Ω_α so that, for each α , $(\varphi_{\alpha,\gamma}, \mathfrak{F}_{\alpha,\gamma})_{\gamma<\alpha}$ forms a martingale and $\varphi_{\alpha,\gamma}(P_\alpha) = \mu_\gamma$. On $\Omega_1 = C$, let $P_1 = \mu_0$ so that $\varphi_{1,0}(P_1) = \mu_0$. Suppose $\alpha > 1$ and P_β has been defined for all $\beta < \alpha$.

First, suppose α is a limit ordinal. Then Ω_α is the inverse limit of $(\Omega_\beta)_{\beta<\alpha}$ and the measures P_β are consistent, so there is a unique extension P_α to Ω_α consistent with the P_β .

Next, suppose $\alpha = \beta + 1$. Then $(\varphi_{\beta,\gamma})_{\gamma<\beta}$ is a martingale, hence converges, say to $\psi_\beta : \Omega_\beta \to C$. (If $\beta = \beta' + 1$, then of course $\psi_\beta = \varphi_{\beta,\beta'}$.) Now the measure $\nu_\beta = \psi_\beta(P_\beta)$ is the least upper bound of $(\mu_\gamma)_{\gamma<\beta}$, so $\nu_\beta \prec \mu_\beta$. Choose a ν_β-dilation T_β so that $T_\beta(\nu_\beta) = \mu_\beta$. Define P_α on $\Omega_\alpha = \Omega_\beta \times C$ by:

$$P_\alpha(A) = \int_C T_\beta(x)(A_\omega) d\nu_\beta(x) \quad ,$$

where $A_\omega = \{x \in C : (\omega,x) \in A\}$ for $\omega \in \Omega_\beta$. Thus $\varphi_{\alpha,\beta}(P_\alpha) = \mu_\beta$ and $E[\varphi_{\alpha,\beta} | \mathfrak{F}_{\alpha,\beta'}] = \varphi_{\alpha,\beta'}$ for $\beta' \leq \beta$. This completes the inductive definition of $(P_\alpha)_{\alpha\leq\xi}$.

As before, $(\varphi_{\xi,\gamma})_{\gamma<\xi}$ is a martingale, hence converges, say to $\psi_\xi : \Omega_\xi \to C$. The measure $\nu_\xi = \psi_\xi(P_\xi)$ is the least upper bound of $(\mu_\gamma)_{\gamma<\xi}$. □

If the condition (b') or (c') is satisfied, then (accoring to Zorn's Lemma) for every $\mu \in \mathcal{P}_t(C)$, there is a maximal $\lambda \in \mathcal{P}_t(C)$ with $\lambda \succ \mu$. This fact is relevant in Choquet-type representation theorems: see [16, p. 25], [6, p. 157].

Next is a result giving conditions under which every maximal measure is concentrated on the extreme points in the sense that $\mu(B) = 1$ for every Borel set $B \supseteq \text{ex } C$. (This is not true in general, even if C is a closed bounded convex set in a Banach space: see [6, p. 159].) Of course, part (iii) is weaker then the known Bishop-de Leeuw Theorem.

2.5 PROPOSITION. *Let* C *be a bounded convex subset of a locally convex space. Suppose either*

(i) C *is analytic; or*

(ii) C *is completely metrizable and* $r: \mathcal{P}_t(C) \to C$ *is open; or*

(iii) C *is compact and* r *is open.*

Then if $\mu \in \mathcal{P}_t(C)$ *is maximal, we have* $\mu(B) = 1$ *for every Borel set* $B \supseteq \text{ex } C$.

Proof. We will prove the contrapositive. Suppose $\mu(B) < 1$ for some Borel set $B \supseteq C$. Then we may assume without loss of generality that there is a compact set $K \subseteq C\setminus\text{ex } C$ with $\mu(K) = 1$. In each of the three cases, C is t-convex, so $r: \mathcal{P}_t(C) \to C$ is defined. Let $\mathcal{R} = \{\varepsilon_x : x \in C\}$. Now $\mathcal{R}$ is closed in $\mathcal{P}_t(C)$ and, in case (i) r is continuous on $\mathcal{P}_t(C)\setminus\mathcal{R}$, which is analytic; in case (ii) r is continuous and open on $\mathcal{P}_t(C)\setminus\mathcal{R}$ which is completely metrizable; in case (iii) r is continuous and open on $\mathcal{P}_t(C)\setminus\mathcal{R}$ which is locally compact. Then (i) by the von Neumann selection theorem or (ii) (iii) by [7], there is a measurable weak section $T: C\setminus\text{ex } C \to \mathcal{P}_t(C)\setminus\mathcal{R}$, i.e. $r \circ T$ is weakly equivalent to the identity on $C\setminus\text{ex } C$. Define $T(x) = \varepsilon_x$ for $x \in \text{ex } C$, so that T is a μ-dilation. We claim that $T(\mu) \neq \mu$; this will show that μ is not maximal.

Now $T \in L^0$, so there is a compact set $\mathcal{D} \subseteq \mathcal{P}_t(C)\setminus\mathcal{R}$ with $\mathcal{D} \subseteq T(K)$ and $\mu(T^{-1}(\mathcal{D}) \cap K) > 0$. But $\mathcal{D}$ is compact, $\mathcal{R}$ is closed, and $\mathcal{D} \cap \mathcal{R} = \emptyset$, so there is a continuous function $h: \mathcal{P}_t(C) \to \mathbb{R}$ such that $h = 0$ on $\mathcal{D}$, $h = 1$ on $\mathcal{R}$ and $0 \leq h \leq 1$. Now $\mu\{x: h(T(x)) \neq h(\varepsilon_x)\} > 0$. But the topology of $\mathcal{P}_t(C)$ is generated by the set of maps $\nu \mapsto \int f \, d\nu$, where f is bounded continuous convex (Proposition 2.1). Thus there is a bounded continuous convex function f such that $\mu\{x: \langle T(x), f\rangle \neq \langle \varepsilon_x, f\rangle\} > 0$. Thus $\langle T(\mu), f\rangle = \int \langle T(x), f\rangle d\mu(x) > \langle \mu, f\rangle$. Therefore $T(\mu) \neq \mu$, and thus μ is not maximal. $\square$

Condition (iii) is studied by R. C. O'Brien in [15].

Let $(\Omega, \mathcal{F}, P)$ be a probability space and C a bounded convex set. Write $\mathcal{F}^+ = \{A \in \mathcal{F}: P(A) > 0\}$. A function $\eta: \mathcal{F}^+ \to C$ is an <u>averaged measure</u> provided

$$\eta(A \cup B) = \frac{P(A)}{P(A \cup B)} \eta(A) + \frac{P(B)}{P(A \cup B)} \eta(B)$$

for disjoint $A, B \in \mathfrak{F}^+$. Clearly, if m is a measure with average range in C, then m/P is an averaged measure with values in C and conversely (countable additivity of $m = P \cdot \eta$ follows from the boundedness of C).

2.6 PROPOSITION. Let C', C be bounded convex sets in locally convex spaces. Suppose $u: C' \to C$ is continuous, bijective and affine. If C' has the Radon-Nikodym property, then C has the Radon-Nikodym property.

Proof. Let $(\Omega, \mathfrak{F}, P)$ be a probability space, m a measure with average range in C. Then $\eta = m/P$ is an averaged measure in C. Define $\eta': \mathfrak{F}^+ \to C'$ by $\eta'(A) = u^{-1}(\eta(A))$. Since u is bijective and affine, η' is an averaged measure in C'. Since C' has the Radon-Nikodym property, there is $\varphi' \in L^0(\Omega, \mathfrak{F}, P; C')$ with $\eta'(A) = P(A)^{-1} \int_A \varphi' \, dP$ for all $A \in \mathfrak{F}^+$. Let $\varphi = u \circ \varphi'$. Now u is continuous, so $\varphi \in L^0(\Omega, \mathfrak{F}, P; C)$. Also u is continuous and affine, so $\eta(A) = u(\eta'(A)) = P(A)^{-1} \int_A u \circ \varphi \, dP = P(A)^{-1} \int_A \varphi \, dP$. Thus C has the Radon-Nikodym property. □

2.7 COROLLARY. Let C be a bounded convex set in a locally convex space. Write $\mathrm{ex}\, C$ for the set of extreme points of C. Suppose that

(1) $\mathrm{ex}\, C$ is relatively t-convex in C; i.e. for every $\mu \in \mathcal{P}_t(\mathrm{ex}\, C)$, there exists $r(\mu) \in C$;

(2) for every $x \in C$; there is a unique $\mu \in \mathcal{P}_t(\mathrm{ex}\, C)$ with $r(\mu) = x$.

Then C has the Radon-Nikodym property.

Proof. First, $\mathcal{P}_t(\mathrm{ex}\, C)$ has the Radon-Nikodym property by Corollary 1.3. The resultant map $r: \mathcal{P}_t(\mathrm{ex}\, C) \to C$ is defined by (1) and bijective by (2); it is always affine and continuous. Thus by Proposition 2.6, C has the Radon-Nikodym property. □

Remarks. (1) For example, if C is a separable closed bounded convex subset of a Banach space, then $\mathrm{ex}\, C$ is universally measurable [2, Prop. 2.1], so $\mathcal{P}_t(\mathrm{ex}\, C)$ can be identified with $\{\mu \in \mathcal{P}_t(C): \mu(\mathrm{ex}\, C) = 1\}$. Thus, if every point of C is represented by a unique measure on $\mathrm{ex}\, C$, then C has the Radon-Nikodym property. This is a (very) partial converse of [5].

(2) If C is a separable closed bounded subset of a Banach space and C is a (noncompact) simplex, does it follow that a point of C can have at most one representing measure on $\mathrm{ex}\, C$? (The Radon-Nikodym property is not postulated, cf. [3, Theorem 1.1].)

(3) If C is a nonseparable closed bounded convex subset of a Hilbert space, the set of maximal measures on C need not have the Radon-Nikodym property (the example in [6, p. 159] exhibits this behavior), so Proposition 2.6 will not apply in this case.

Note. After this paper was written, H. von Weizsäcker kindly gave me a copy of his paper "Einige masstheoretische Formen der Sätze von Krein-Milman und Choquet". It has considerable overlap with the present paper. Among many other things, von Weizsäcker gives an example of a completely regular space T for which $\mathcal{P}_t(T)$ fails the martingale convergence property. (See von Weizsäcker's paper in this volume.)

References

1. G. Birkhoff, Lattice Theory, Third ed., American Mathematical Society, Providence, R.I., 1967.
2. R. D. Bourgin, Barycenters of measures on certain noncompact convex sets, Trans. Amer. Math. Soc. 154 (1971) 323-340.
3. R. D. Bourgin and G. A. Edgar, Noncompact simplexes in spaces with the Radon-Nikodym property, J. Functional Analysis 23 (1976) 162-176.
4. G. A. Edgar, Disintegration of measures and the vector-valued Radon-Nikodym theorem, Duke Math. J. 42 (1974) 447-450.
5. G. A. Edgar, A noncompact Choquet theorem, Proc. Amer. Math. Soc. 49 (1975) 354-358.
6. G. A. Edgar, Extremal integral representations, J. Functional Analysis 23 (1976) 145-161.
7. G. A. Edgar, Measurable weak sections, Illinois J. Math. 20 (1976) 630-646.
8. D. H. Fremlin and I. Pryce, Semiextremal sets and measure representation, Proc. London Math. Soc. (3) 29 (1974) 502-520.
9. A. Goldman, Measures cylindriques, measures vectorielles et questions de concentration cylindrique, Pacific J. Math. (to appear).
10. J. Hoffmann-Jørgensen, Weak compactness and tightness of subsets of M(X), Math. Scand. 31 (1972) 127-150.
11. J. Hoffmann-Jørgensen, The strong law of large numbers and the central limit theorem in Banach spaces, Aarhus Universitet, Various Publications Series no. 24, pp. 74-99.
12. J. Hoffmann-Jørgensen and G. Pisier, The law of large numbers and the central limit theorem in Banach spaces, Ann. of Prob. 4 (1976) 587-599.
13. J. Kupka, Radon-Nikodym theorems for vector-valued measures, Trans. Amer. Math. Soc. 169 (1972) 197-217.
14. L. LeCam, Convergence in distribution of stochastic processes, Univ. of Calif. Publ. in Stat. 2 (1957) 207-236.
15. R. C. O'Brien, On the openness of the barycentre map, Math. Ann. 223 (1976) 207-212.
16. R. R. Phelps, Lectures on Choquet's Theorem, Van Nostrand, New York, 1966.
17. V. Strassen, The existence of probability measures with given marginals, Ann. of Math. Stat. 36 (1965) 423-439.
18. V. S. Varadarajan, Measures on topological spaces, Amer. Math. Soc. Transl. (2) 48 (1965) 161-228.

19. H. von Weizsäcker, Der Satz von Choquet-Bishop-de Leeuw für konvexe nicht kompakte Mengen straffer Masse über beliebigen Grundräumen, Math. Z. 142 (1975) 161-165.

The Ohio State University
Columbus, Ohio 43210 U.S.A.

On the Radon-Nikodym-Property, and related topics in locally convex spaces

by

L.EGGHE
(L.U.C.-Hasselt)

ABSTRACT

We introduce $L^1_X(\mu)$, the space of classes of X-valued μ-integrable functions used by Saab, which is an extension of the space of classes of Bochner-integrable functions, in Banach spaces. X denotes here a sequentially complete locally convex space.

We give examples of spaces which are dentable, σ-dentable, having the Radon-Nikodym-Property, or having the Bishop-Phelps-Property, by proving some projective limit results.

We also prove the following theorem : The following implications are valid :

$$(i) \Leftrightarrow (ii) \Rightarrow (iii) \Leftrightarrow (iv) \Leftrightarrow (v)$$

(i) X has the Radon-Nikodym-Property.
(ii) Every uniformly bounded martingale is L^1_X-convergent.
(iii) Every uniformly bounded martingale is L^1_X-Cauchy.
(iv) Every uniformly bounded and finitely generated martingale is L^1_X-Cauchy.
(v) X is σ-dentable.

So we have the equivalency of (i) through (v) for quasi-complete (BM)-spaces.

§ 1. Introduction, terminology and notation.

A non-empty subset B of a locally convex space (l.c.s.) E (always supposed to be over the reals) is called dentable, if for every neighborhood (nbhd) V of 0, there exists a point x in B such that

$$x \notin \overline{\text{con}}\ (B \setminus (x+V))$$

I thank Dr. J.A. Van Casteren for helping me preparing this paper.

where $\overline{\text{con}}$ denotes the closed convex hull. If the same x can be chosen, for every V in the above definition, we call x a denting point of B. We call E dentable if every bounded subset of E is dentable. When we use $\sigma(A)$ instead of $\overline{\text{con}}(A)$, we get the corresponding definitions for σ-dentability, and σ-denting point ; Here

$$\sigma(A) = \left\{ \sum_{n=1}^{\infty} \lambda_n x_n \,\|\, x_n \in A, \ \sum_{n=1}^{\infty} \lambda_n = 1, \ \sum_{n=1}^{\infty} \lambda_n x_n \text{ convergent}, \lambda_n > 0 \right\}$$

We use the following integral [7] : Let E be a sequentially complete l.c.s., and (Ω,Σ,μ) be a finite complete positive measure space. A function $f:\Omega\to X$ is said to be μ-integrable, if there exists a sequence $(f_n)_{n=1}^{\infty}$ of simple functions such that:

(i) $\lim_n f_n(\omega) = f(\omega)$, μ-a.e.
(ii)For every continuous seminorm p on E : $\lim_n \int_\Omega p(f_n(\omega) - f(\omega))\, d\mu(\omega) = 0$.

Put $\int_A f d\mu = \lim_n \int_A f_n\, d\mu$, $\forall A \in \Sigma$. This limit exists and is in E. Denote $L_E^1(\mu,\Sigma)$ as the space of classes $[f]$ of μ-integrable functions, where $g\in[f]$ iff $f=g$, μ a.e.. Put

$$q(f) = \int_\Omega p(f)\, d\,\mu,$$

where p is any continuous seminorm on E. The topology on L_E^1 considered is these, generated by the q.

Let B be a closed bounded subset of E. We say that B has the Radon-Nikodym-Property (RNP), if for every positive finite measure space (Ω,Σ,μ), and every vector-measure $m:\Sigma\to E$, with

$$A_m(\Sigma,\mu) = \left\{ \frac{m(A)}{\mu(A)} \,\|\, A \in \Sigma,\ \mu(A) > 0 \right\}$$

contained in B, there is a μ-integrable function $f:\Omega \to E$, such that

$$m(A) = \int_A f\, d\mu\ , \quad \forall A \in \Sigma$$

We say that E has the (RNP) if each closed bounded convex subset of E has the (RNP).

Let (Ω,Σ,μ) be a positive finite complete measure space. A net $(x_i, \Sigma_i)_{i\in I}$, where I is a directed set, is called a E-valued martingale, if :

(i) $i \leqslant j$ implies $\Sigma_i \subset \Sigma_j$

(ii) Every Σ_i is a sub-σ-algebra of Σ

(iii) Every x_i is in $L^1_X(\mu, \Sigma_i)$

(iv) $\forall\, i \in I$, $\forall\, A \in \Sigma_i$, we have, $\forall\, j \geqslant i : \int_A x_j \, d\mu = \int_A x_i \, d\mu$

Let E be a Banach space. E is said to have the Bishop-Phelps-Property (BPP), if for any Banach space F, any closed bounded absolutely convex subset B in E, and any continuous linear operator T from E into F, there is a sequence (T_n) of continuous linear operators from E into F, such that $\lim \|T_n - T\| = 0$, and each T_n attains its sup-norm on B : i.e. : There exists an x_n in B such that :

$$\sup_{x \in B} \|T_n x\| = \|T_n x_n\|$$

Following Saab [7] , a (BM)-space is a l.c.s. in which every bounded subset is metrizable. In a quasi-complete (BM)-space, Saab proves the equivalency of dentability, σ-dentability, (RNP).

§ 2. Projective limits of dentable spaces.

In giving examples of l.c.s. being dentable, σ-dentable or having the (RNP), the following permanency result is useful :

<u>THEOREM 1</u> : The projective limit of a family of l.c.s., a cofinal system of them being dentable, is dentable.

<u>Proof</u> : Let $E = \varprojlim g_{\alpha\beta} E_\beta$, with $(E_\alpha)_{\alpha\in\Lambda}$ the family of l.c.s., and Λ a directed set. Supposing a bounded set B of E not dentable, there is a nbhd B of 0, which we can take of the form

$$V = \Big(\prod_{i=1}^{n} V_{\alpha_i} \times \prod_{\substack{\alpha \neq \alpha_i \\ i=1,\dots,n}} E_\alpha \Big) \cap E$$

such that $x \in B$ implies $x \in \overline{\text{con}}\,(B \setminus (x+V))$. Choose $\beta\in\Lambda$ such that $\beta \geqslant \alpha_j$, $j=1,\dots,n$, and E_β is dentable. Denote by p_{α_j} the gauge of V_{α_j}. For every p_{α_j} there exists a continuous seminorm $p_\beta^{(j)}$ on E_β, and a number $c_{\alpha_j\beta} \in\,]0, +\infty[$ such that

$$p_{\alpha_j}(g_{\alpha_j\beta}(x_\beta)) \leqslant c_{\alpha_j\beta}\, p_\beta^{(j)}(x_\beta)$$

for all $x_\beta \in E_\beta$. Taking $p_\beta = \max_{j=1,\dots,n} c_{\alpha_j\beta}\, p_\beta^{(j)}$,

$$V_\beta = \left\{ x_\beta \in E_\beta \, \| p_\beta(x_\beta) \leqslant 1 \right\}, \quad U_x = x + \left[V \cap [\prod_{\alpha \neq \beta} E_\alpha \times V_\beta] \right]$$

we have $x_\beta = g_\beta(x) \in \overline{\text{con}}\,(g_\beta(B) \setminus (x_\beta + V_\beta))$, which is by the boundedness of $g_\beta(B)$ a contradiction with the fact that E_β is dentable. □

Corollary 1 : Every nuclear space is dentable.

Corollary 2 : Every quasi-complete nuclear (BM)-space has (RNP).

Proof : Use [7] and corollary 1. □

Example : Any separable Hilbert space with the weak topology is an example of a quasi-complete nuclear (BM)-space, not being metrizable.

Proposition : A countable product of quasi-complete l.c.s. has the (RNP) if every factor space has the (RNP).

Proof : This is straightforward . □

Corollary 3 : Every product of quasi-complete (BM)-spaces with the (RNP) is dentable.

Hence from [7] we also have :

Corollary 4 : The countable projective limit of (RNP) quasi-complete (BM)-spaces is quasi-complete, (BM) and has the (RNP).

Corollary 5 : A countable product of quasi-complete (BM)-spaces has the (RNP) iff each factor space has the (RNP).

By the same method as in theorem 1 we can prove :

THEOREM 2 : Let E be a projective limit of l.c.s. :

$$E = \varprojlim g_{\alpha\beta} \, E_\beta$$

Let B be a subset of E. If $g_\alpha(B)$ is dentable in E_α for a cofinal system of α , then B is dentable in E.

Let dent(B) denote the set of the denting points of B.

THEOREM 3 : When $E = \varprojlim g_{\alpha\beta} E_\beta$, $B \subset E$, and $x = (x_\alpha) \in E$, with $x_\alpha = g_\alpha(x) \in \text{dent}(g_\alpha(B))$ for a cofinal system of α, then $x \in \text{dent}(B)$.

Again the same argument gives the result for σ-dentability, in theorems 1,2,3.

Corollary 6 : (i) A semi-reflexive l.c.s. E is dentable.

(ii) A quasi-complete semi-reflexive (BM)-space has the (RNP).

Proof : (i) follows easily from theorem 2, and (ii) from [7] and (i). □

An interesting example is the space $\mathcal{L}_s(\mathcal{H})$, where $\mathcal{H}$ is a separable Hilbertspace, and s denotes the strong operator topology. We have (by corollary 3)

THEOREM 4 : If $\mathcal{H}$ is a separable Hilbert space, then $\mathcal{L}_s(\mathcal{H})$ has the (RNP).

§3. The Bishop-Phelps-Property in l.c.s..

Definition : Let E be a l.c.s.. We say that E has the Bishop-Phelps- property (BPP), if for every Banach space F, every T in $\mathcal{L}(E,F)$, and every closed bounded and absolutely convex set B in E, there exists a sequence $(T_n)_{n=1}^{\infty}$ in $\mathcal{L}(E,F)$, such that $T_n \to T$ for $n \to \infty$, uniformly on the bounded subsets of E, and such that, for every n in N, there is an x_n in B, such that

$$\sup_{x\in B} \|T_n x\| = \| T_n x_n\|$$

The following lemma is easily seen :

Lemma : Let E be the projective limit of a family of l.c.s. $(E_\alpha)_{\alpha\in\Lambda}$ with maps $g_{\alpha\beta}$ $(\alpha \leqslant \beta)$ and g_α. Let F be an arbitrary Banach space and T arbitrary in $\mathcal{L}(E,F)$. Then there is an $\alpha_0 \in \Lambda$ such that $\ker T \supset \ker g_\alpha$, for every $\alpha \geqslant \alpha_0$.

Definition : We call a l.c.s. E the proper projective limit of the Banach spaces $(E_\alpha)_{\alpha\in\Lambda}$ (w.r.t. mappings $g_\alpha: E \to E_\alpha$ and $g_{\alpha\beta} : E_\beta \to E_\alpha$), if E is the projective limit of these $(E_\alpha, g_\alpha, g_{\alpha\beta})$, and the maps g_α transform every closed bounded absolutely convex set into a closed set.

A complete semi-reflexive l.c.s. is always a proper projective limit of Banach spaces.

THEOREM 5 : A proper projective limit E of a family of Banach spaces $(E_\alpha)_{\alpha\in\Lambda}$, a cofinal system of them having the (BPP) has the (BPP).

Proof : Let F, T, and α as in the lemma, and E_α having the (BPP). Let B be a fixed but arbitrary closed bounded and absolutely convex subset of E. Denote $g_\alpha(B)$ as B_α. Put $S(x_\alpha) = Tx$ $(x_\alpha = g_\alpha(x))$. By the lemma we see that the definition of S makes sense, and it is easy to see that S is an element of $\mathcal{L}(g_\alpha(E), F)$. Extend S to $\hat{S} \in \mathcal{L}(\overline{g_\alpha(E)}, F)$ where the closure is taken in E_α. Since E_α has the (BPP), we see by [1], that $\overline{g_\alpha(E)}$ has the (BPP). So there exists a sequence $\hat{S}_n \in \mathcal{L}(\overline{g_\alpha(E)}, F)$, together with a sequence $x^n_\alpha \in B_\alpha$, such that $\lim_n \|\hat{S}_n - \hat{S}\| = 0$, and such that

$$\|\hat{S}_n(x^n_\alpha)\| = \sup_{x_\alpha \in B_\alpha} \|\hat{S}_n(x_\alpha)\| ,$$

for every n in N. (since B_α is closed). Put $S_n = \hat{S}_n | g_\alpha(E)$; then $S_n \in \mathcal{L}(g_\alpha(E), F)$, $\lim_n \|S_n - S\| = 0$ and

$$\|S_n(x^n_\alpha)\| = \sup_{x_\alpha \in B_\alpha} \|S_n x_\alpha\|$$

Put $T_n = S_n \circ g_\alpha$. Then it is easy to see that T_n suffices all the requirements of the definition (for every n in N we take $x^n \in B$ such that $g_\alpha(x^n) = x^n_\alpha$). □

Corollary : Every projective limit of reflexive Banach spaces has the (BPP).

Remarks : 1. From the result in [1] we only used the fact that the (BPP) is hereditary for closed subspaces. It is worth remarking that a direct proof of this fact is not known, nor is the corresponding question in l.c.s..

2. By the fact that each factor space is complemented in a product, we have that if a product of Banach spaces has (BPP), then each factor space has (BPP).

§ 4. (RNP), σ-dentability and martingale-convergence properties.

We are now dealing with the following well-known theorem in Banach spaces, [2] , [6] .

THEOREM : Let E be a Banach space. The following assertions are equivalent :

(i) E has (RNP).

(ii) Every uniformly bounded martingale $(x_n, \Sigma_n)_{n=1}^\infty$ is L^1_E - convergent.

(iii) E is dentable.

(iv) E is σ-dentable.

The equivalency of (i), (iii) and (iv) is also known in quasi-complete (BM)-spaces ([7]). In our case, the space $L^1_E(\mu, \Sigma)$ is in general not complete, so that we

might get some Cauchy-results when (ii) is relied to (iii) or (iv). However, the existence of a Radon-Nikodym-derivative for a vectormeasure (w.r.t. a positive finite measure) supposes a certain completeness-property for $L^1_E(\mu)$.
In fact we prove the following theorem :

THEOREM 6 : Let E be a sequentially complete l.c.s.. Then the following is valid :

$$(i) \Leftrightarrow (ii) \Rightarrow (iii) \Leftrightarrow (iv) \Leftrightarrow (v)$$

(i) E has the (RNP).
(ii) Every uniformly bounded martingale $(x_i, \Sigma_i)_{i \in I}$ is L^1_E-convergent. (I onbitrary)
(iii) Every uniformly bounded martingale $(x_n, \Sigma_n)_{n=1}^{\infty}$ is L^1_E-Cauchy.
(iv) Every uniformly bounded and finitely generated martingale $(x_n, \Sigma_n)_{n=1}^{\infty}$, is L^1_E-Cauchy.
(v) E is σ-dentable.

The proof is divided into several steps.

A. (i) ⇔ (ii)

We can take the proof of [2] , p.271 - 274, up to minor modifications. We have here the existence of conditional expectations :

$$E(., \Sigma_1): L^1_E(\mu, \Sigma) \rightarrow L^1_E(\mu, \Sigma_1)$$

where $\Sigma_1 \subset \Sigma$. For step functions this is seen as in [2] . For $x \in L^1_E(\mu, \Sigma)$ and (x_n) stepfunctions such that $\lim_n x_n = x$,μ-a.e. and $\lim_n q(x_n - x) = 0$ for every continuous seminorm q on $L^1_E(\mu, \Sigma)$, we see immediately that $E(x, \Sigma_1) = \lim_n E(x_n, \Sigma_1)$ exists μ / Σ_1 - a.e., and that $(E(x_n, \Sigma_1))_{n=1}^{\infty}$ is L^1_E-Cauchy. Hence $E(x, \Sigma_1) \in L^1_E$.

Furthermore $E(., \Sigma_1)$ is also seen to be continuous and linear.

Since (ii) ⇒ (iii) ⇒ (iv) is trivial, we come to :

B. (iv) ⇒ (v)

For this we adapt the proof of Huff in [5] to our case. We give only the differences between our construction and that of Huff.
Suppose there is a bounded set $B \subset E$, not σ-dentable, then there is a nbhd V of 0, such that $x \in B$ implies $x \in \sigma(B \setminus (x + V))$. We contruct a σ-algebra $\Sigma \subset B_{[0,1)}$, the

Borel subsets of $[0,1)$, which is generated by $\bigcup_{n=1} \pi_n$, where $\pi_n = \{I^n_{0,1},\dots,I^n_{0,k_n}; I^n_1,\dots,I^n_{i_n}\}$ a partition of $[0,1)$ into intervals of the same form. Put $\mu = \lambda/\Sigma$, with λ the Lebesgue measure on $[0,1)$. We construct $x^n_i \in B$ $(i=1,\dots,i_n)(n \in N)$, and

$$f_n = \sum_{i=1}^{i_n} x^n_i \chi_{I^n_i},$$

such that :

(1) $(\pi_n)_{n=1}^{\infty}$ is increasing. ($\pi_1 \leq \pi_2$ means : every element of π_1 is (μ-a.e.) union of elements of π_2).

(2) $\mu(A) \geq \frac{1}{4}$, where $A = \bigcap_{n=1}^{\infty} [\bigcup_{i=1}^{i_n} I^n_i]$

(3) $p_V(f_n(\omega) - f_{n+1}(\omega)) \geq 1$, $\forall \omega \in A$ (p_V denotes the gauge of V).

(4) $p(\int_E (f_n - f_{n+1})\,d\mu) \leq \frac{1}{2^n} M_p.\mu(E)$, $\forall E \in \Sigma$ (M_p denotes a p-bound vor B, p being

an arbitrary continuous seminorm on E.

(5) $\nu(E) = \lim_n \int_E f_n d\mu$ exists, for every E in Σ. When this should be done, we can

prove inthe same way as in [5] the contradiction.

<u>The Construction</u> : Let p be an arbitrary continuous seminorm on E. Put $\pi_o = \{[0,1)\}$ and $f_o = x^o_1 \chi_{[0,1)}$, with x^o_1 arbitrary in B. Suppose now $\pi_o,\dots,\pi_n$ already constructed, consisting of finite partitions of $[0,1)$, into intervals of the same form, and

$$f_n = \sum_{i=1}^{i_n} x^n_i \chi_{I^n_i}, \quad x^n_i \in B, \quad \pi_n = \{I^n_{0,1} \dots I^n_{0,k_n}; I^n_1,\dots,I^n_{i_n}\}$$

For any $x^n_i \in B$, there exists

(i) a sequence $y^{(i)}_1,\dots, y^{(j)}_p,\dots$ in $B \setminus (x^n_i + V)$

(ii) a sequence $\alpha^{(i)}_1, \dots, \alpha^{(i)}_p, \dots$ in $[0,1]$

such that

$$x^n_i = \sum_{j=1}^{\infty} \alpha^{(i)}_j y^{(i)}_j, \quad \sum_{j=1}^{\infty} \alpha^{(i)}_j = 1$$

for every $i = 1,\dots, i_n$. Choose $j^{(i)}_{n+1} \in N$ such that :

$$\sum_{j=j^{(i)}_{n+1}+1}^{\infty} \alpha^{(i)}_j \leq \frac{1}{2^n}$$

Hence :

$$p(x_i^n - \sum_{j=1}^{j_{n+1}^{(i)}} \alpha_j^{(i)} y_j^{(i)}) \leqslant \frac{M_p}{2^n}$$

Divide I_i^n into half-open, disjoint intervals $J_j^{(i)} (j=1,\dots,J_{n+1}^{(i)})$ such that

$$\lambda(J_j^{(i)}) = \alpha_j^{(i)} \lambda(I_i^n)$$

Call

$$J_o^{(i)} = I_i^n \setminus \bigcup_{j=1}^{j_{n+1}^{(i)}} J_j^{(i)}$$

$$\pi_{n+1} = \left\{ J_j^{(i)} \parallel j = 0,1,\dots,j_{n+1}^{(i)} \; ; \; i = 1,\dots,i_n \right\} \cup \left\{ I_{0,1}^n,\dots,I_{0,k_n}^n \right\}$$

and

$$f_{n+1} = \sum_{i=1}^{i_n} \sum_{j=1}^{j_{n+1}^{(i)}} y_j^{(i)} \chi_{J_j^{(i)}}$$

Denote $\left\{ I_{0,1}^n,\dots,I_{0,k_n}^n \right\} \cup \left\{ J_0^{(1)},\dots,J_0^{(i_n)} \right\} = \left\{ I_{0,1}^{n+1},\dots,I_{0,k_{n+1}}^{n+1} \right\}$

and $\left\{ J_j^{(i)} \parallel j=1,\dots,j_{n+1}^{(i)}; \; i=1,\dots,i_n \right\} = \left\{ I_1^{n+1},\dots,I_{i_{n+1}}^{n+1} \right\}$

Then $\pi_{n+1} = \left\{ I_{0,1}^{n+1},\dots,I_{0,k_{n+1}}^{n+1} \; ; \; I_1^{n+1},\dots,I_{i_{n+1}}^{n+1} \right\}$

Put $A = \bigcap_{n=1}^{\infty} [\bigcup_{i=1}^{i_n} I_i^n]$, Σ the σ-algebra, generated by $\bigcup_{n=1}^{\infty} \pi_n$, and $\mu = \lambda | \Sigma$.

Then all the conditions for the construction are verified as in Huff's proof, using here the fact that f_n is 0 on $I_{0,1}^n,\dots,I_{0,k_n}^n$.

Corollary : Let E be a quasi-complete (BM)-space. Then all the assertions in the theorem are equivalent (and equivalent with dentability).

We return to the case when E is a sequentially complete l.c.s..

C. (v) $\Rightarrow$ (iv)

Let $(x_n, \Sigma_n)_{n=1}^{\infty}$ be a finitely generated martingale. So every x_n is of the form :

$$x_n = \sum_{i=1}^{i_n} a_i^{(n)} \chi_{A_i^{(n)}} , \quad a_i^{(n)} \in E, \quad A_i^{(n)} \in \Sigma_n.$$

Put $F_n(A) = \int_A x_n \, d\mu$, $\forall A \in \Sigma_n$, then it is easily seen that $F(A) = \lim_n F_n(A)$ exists, for every A in Σ, the σ-algebra generated by the Σ_n.

Put :

$$y = \sum_{A \in \pi} \frac{F(A)}{\mu(A)} \chi_A$$

with π an arbitrary finite partition of Ω in elements of Σ. We call π_n the finest partition of Ω in elements of Σ_n. It is trivial that $y_{\pi_n} = x_n$, $\forall n \in \mathbb{N}$. So we have to prove that $(y_{\pi_n})_{n=1}^{\infty}$ is L^1_E-Cauchy. From a trivial adaptation of a part of the proof of the theorem of Rieffel, to our case, (see [2]), we see that $(y_\pi)_{\pi \in \Pi}$ is L^1_E-Cauchy. (where Π denotes the set of all finite partitions of Ω in elements of Σ). Hence for every continuous seminorm p on E, there is a $\pi_o \in \Pi$, such that for every $\pi \geqslant \pi_o$:

$$q(y_\pi - y_{\pi_o}) \leqslant \frac{1}{4} .$$

Let $\pi_o = \{A_1, \dots, A_n\}$. We can construct

$$\left\{A'_1, \dots, A'_n, \Omega \setminus \bigcup_{i=1}^{n} A'_i\right\}$$

in $\bigcup_{n=1}^{\infty} \Sigma_n$, such that $\mu(A_i \, \Delta \, A'_i) < \dfrac{1}{24.n.M_p}$

(where Δ denotes the symmetric difference, $i=1,\dots,n$, and M_p is a p-bound for the uniformly bounded martingale). Put

$$A''_1 = A'_1 \, , \; A''_i = A'_i \setminus \bigcup_{j=1}^{i-1} A'_j \; (i = 2,\dots,n) \, , \; A''_{n+1} = \Omega \setminus \bigcup_{i=1}^{n} A''_i$$

and $\pi''_o = \left\{A''_1, \dots, A''_n, A''_{n+1}\right\}$. Let π' be any refinement of π''_o :

$$\pi' = \left\{B_{1,1},\dots,B_{1,p_1} \, ; \, \dots \, ; \, B_{n,1},\dots,B_{n,p_n} \, ; \, B_{n+1,1},\dots,B_{n+1,p_{n+1}}\right\}$$

Choose $\pi'' \geqslant \pi'$ and π_o, in Π. We consider three parts in π'' :

(I) Those sets $B_{i,j}$ of π' which can also be taken in π'' : i.e. which are already part of one A_k. This part cancels in $y_{\pi'} - y_{\pi''}$.

(II) Those $B_{i,j}$ in π' $(1 \leqslant i \leqslant n)$ which are in more than one A_k. As sets in π'' we have of course to choose $B_{i,j} \cap A_k$ $(k=1,\dots,n)$.

(III) The same as (II), for $i = n+1$.

We have :

$q(y_{\pi'} - y_{\pi''})$

$$\leq \sum_{(II)} p\left(\frac{F(B_{i,j})}{\mu(B_{i,j})} - \frac{F(B_{i,j} \cap A_i)}{\mu(B_{i,j} \cap A_i)} \right) \mu(B_{i,j} \cap A_i)$$

$$+ \sum_{(II)} \sum_{k \neq i} p\left(\frac{F(B_{i,j})}{\mu(B_{i,j})} - \frac{F(B_{i,j} \cap A_k)}{\mu(B_{i,j} \cap A_k)} \right) \mu(B_{i,j} \cap A_k)$$

$$+ \sum_{(III)} \sum_{k=1}^{n} p\left(\frac{F(B_{n+1,j})}{\mu(B_{n+1,j})} - \frac{F(B_{n+1,j} \cap A_k)}{\mu(B_{n+1,j} \cap A_k)} \right) \mu(B_{n+1,j} \cap A_k)$$

$= (1) + (2) + (3)$

We easily see

$$p\left(\frac{F(B_{i,j})}{\mu(B_{i,j})} - \frac{F(B_{i,j} \cap A_i)}{\mu(B_{i,j} \cap A_i)} \right)$$

$$\leq \frac{|F|_p(B_{i,j} \,\Delta\, (B_{i,j} \cap A_i))}{\mu(B_{i,j})} + p\left(\frac{F(B_{i,j} \cap A_i)}{\mu(B_{i,j} \cap A_i)} \right) . \frac{\mu(B_{i,j} \,\Delta\, (B_{i,j} \cap A_i))}{\mu(B_{i,j})}$$

(where $|F|_p$ denotes the p-variation of F)

So $(1) < \frac{1}{12}$

Furthermore we have immediately :

$(2) < \frac{1}{12}$ (since $\bigcup_{k \neq i} A''_i \cap A_k \subset A''_i \setminus A_i$)

and $(3) < \frac{1}{12}$

Hence : $q(y_{\pi'} - y_{\pi''}) < \frac{1}{4}$

Since there is a $n_o \in N$ such that $\pi_{n_o} \geq \pi''_o$, we see immediately that

$$q(y_{\pi_{n_o}} - y_{\pi_o}) < \frac{1}{2}$$

Hence $(y_{\pi_n})_{n=1}^{\infty}$ must be L^1_E - Cauchy.

D. <u>(iv) ⇒ (iii)</u> : We first prove a lemma :

<u>Lemma</u> : Let $x : \Omega \to E$ be (μ, Σ)-integrable. Then there is a sequence $(\pi_n)_{n=1}^{\infty}$ of finite partitions of Ω in elements of Σ , such that $\pi_{n+1} \geq \pi_n$, $\forall\, n \in N$,

and such that $x = L^1_E - \lim_n x_n$, with

$$x_n = \sum_{A \in \pi_n} \frac{\int_A x \, d\mu}{\mu(A)} \chi_A .$$

Proof : Let $(f_n)_{n=1}$ be a sequence of stepfunctions such that $\lim_n q(f_n - x) = 0$, for every continuous seminorm on L^1_E. Suppose $m_q \in N$ is such that

$$q(f_{m_q+k} - x) < \frac{1}{2}$$

for every $k \in N$. We do the following inductive construction :

Take $f^1_1 = \sum_{i=1}^{p_1} a^1_i \chi_{A^1_i} = f_1$ and put $\Sigma'_1 = \sigma(\{A^1_1, \ldots, A^1_{p_1}\})$ (where σ denotes the (σ-)algebra generated by).

Let $\Sigma'_1, \ldots, \Sigma'_n$ be constructed finite σ-algebras such that $f^n_1, \ldots, f^n_n = f_n$ forms a martingale, and let

$$f^n_i = \sum_{j=1}^{p_i} b(n)^i_j \chi_{A^i_j}$$

Take $f^{n+1}_{n+1} = f_{n+1} = \sum_{j=1}^{p_{n+1}} a^{n+1}_j \chi_{A^{n+1}_j}$, define $f^{n+1}_n = \sum_{i=1}^{p_n} b(n+1)^n_i \chi_{A^n_i}$,

with

$$b(n+1)^n_i = \sum_{j=1}^{p_{n+1}} \frac{\mu(A^{n+1}_j \cap A^n_i)}{\mu(A^n_i)} a^{n+1}_j$$

So for every $A \in \Sigma'_n$: $\int_A f^{n+1}_n \, d\mu = \int_A f^{n+1}_{n+1} \, d\mu$.

Do the same with f^{n+1}_n, now using the sets $A^{(n-1)}_i$ $(i=1,\ldots,p_{n-1})$, so defining f^{n+1}_{n-1}, and so on. Hence we have

$$f^{n+1}_1, \ldots, f^{n+1}_n, f^{n+1}_{n+1} .$$

Put :

$$\Sigma'_{n+1} = \sigma(\bigcup_{i=1}^{n} \Sigma'_i \cup \{A^{n+1}_j \parallel j=1,\ldots,p_{n+1}\}).$$

When $0 < s < k$, it is easily seen, by integrating over the sets $A^{m_q+s}_i$ $(i=1,\ldots,p_{m_q+s})$ that

$$\int_\Omega p(f^{m_q+k}_{m_q+s} - f^{m_q+s}_{m_q+s}) \, d\mu < 1$$

Since $\int_{A^{m_q+s}_i} f^{m_q+k}_{m_q+s} \, d\mu = \int_{A^{m_q+s}_i} f^{m_q+k}_{m_q+k} \, d\mu$, we have

$$\lim_k \int_{A_i^{m_q+s}} f_{m_q+s}^{m_q+k} \, d\mu = \int_{A_i^{m_q+s}} x \, d\mu$$

Define

$$y_{m_q+s} = \Sigma_{i=1}^{p_{m_q+s}} \frac{\int_{A_i^{m_q+s}} x \, d\mu}{\mu(A_i^{m_q+s})} \chi_{A_i^{m_q+s}}$$

Hence $$\lim_k \int_{A_i^{m_q+s}} f_{m_q+s}^{m_q+k} \, d\mu = \int_{A_i^{m_q+s}} y_{m_q+s} \, d\mu$$

So also $$\lim_k q(f_{m_q+s}^{m_q+k} - y_{m_q+s}) = 0,$$

since s is fixed. Hence $q(x - y_{m_q+s}) < \frac{3}{2}$. So L_E^1-$\lim y_n = x$.

Choose π_n to be the finest partition of Ω into elements of Σ_n'. It is trivial that y_n can be seen as a stepfunction over π_n, and that (y_n) is a martingale. Hence necessarily

$$y_n = \sum_{A \in \pi_n} \frac{\int_A x \, d\mu}{\mu(A)} \chi_A .$$

Proof of (iv) ⇒ (iii)

Apply the lemma to every x_n, where (x_n, Σ_n) is the given martingale. It is immediate that we can construct in this way a table of stepfunctions $x_{n,m}$ together with finite partitions $\pi_{n,m}$ which are horizontally and vertically increasing, and so that every subsequence $(x_{n_j,m_j}, \sigma(\pi_{n_j,m_j}))_{j=1}^{\infty}$ forms a martingale. Supposing (x_n) not L_E^1-Cauchy, we get such a sequence $(x_{n_j,m_j})_{j=1}^{\infty}$ which is not L_E^1-Cauchy, contradicting (iv). □

Proposition : Let $(X_n)_{n=1}^{\infty}$ be a sequence of sequentially complete spaces, possessing property (iii) in the theorem above. Then $\prod_{n=1}^{\infty} X_n$ possesses also this property.

Proof : This is straightforward. □

Note : By (iii) ⇔ (v), we have also another proof of the part of the theorem 1 about

σ-dentability.

Corollary : The arbitrary product of sequentially complete σ-dentable l.c.s. is σ-dentable.

Proof : This follows from the proposition and from theorem 1. □

Problem : What is the place of dentability in theorem 6 ?

References

[1] BOURGAIN J. On dentability and the Bishop-Phelps property (to appear in Israel Journal of Math.).

[2] DIESTEL J. Geometry of Banach spaces - Selected Topics. Lecture notes in Math. nr.485,1975, Springer Verlag, Berlin.

[3] DIESTEL J. and UHL J.J.Jr. The Radon-Nikodym theorem for Banachspace-valued measures. Rocky Mountain Math.J.,6, nr.1, 1976, p.1 - 46.

[4] DINCULEANU N. Vector Measures. Pergamon Press , Vol. 95, 1967.

[5] HUFF R.E. Dentability and the Radon-Nikodym-Property. Duke Math. J. 41 (1974) p.111-114.

[6] MAYNARD H. A geometric characterization of Banach spaces possessing the Radon-Nikodym-Property. Trans. AMS, 185, 1973,p. 493 - 500.

[7] SAAB E. On the Radon-Nikodym-Property in locally convex spaces of type (BM). (preprint).

[8] SCHAEFER H.H. Topological Vector Spaces, Springer Verlag, Graduate Texts in Math. 3, 1971 , Berlin.

CURRENT ADDRESS

L.EGGHE
Limburgs Universitair Centrum
Universitaire Campus
B-3610 DIEPENBEEK (Belgium)

RELATIONS ENTRE LES PROPRIETES DE MESURABILITE UNIVERSELLE POUR UN ESPACE TOPOLOGIQUE T ET LA PROPRIETE DE RADON-NIKODYM POUR LE CONE POSITIF DES MESURES DE RADON (RESP. DE BAIRE) SUR T

par André Goldman
Dept. de Mathematiques,
Univ. Claude Bernard (Lyon I),
69 Villeurbanne, Lyon, France.

INTRODUCTION.

Dans ce travail, nous reprenons en partie des résultats sur la propriété de Radon-Nikodym, publiés au Pacific Journal of Mathematics (5). Ces questions sont en relation étroite avec des problèmes de désintégration de mesures définies sur un espace produit.

NOTATIONS.

On désignera par $\vec{m} : \Sigma \to E$ une mesure vectorielle à valeurs dans un elc E, absolument continue par rapport à une probabilité μ sur l'espace mesurable (Ω,Σ). On supposera toujours que l'ensemble $S_\Omega = \{\frac{\vec{m}(A)}{\mu(A)} ; A \in \Sigma, \mu(A) > 0\}$ est borné dans E, ce qui, dans le cas où E est un espace normé, correspond au fait que la mesure $\vec{m}$ est à variation bornée. Nous appellerons densité faible toute fonction vectorielle $\vec{f} : \Omega \to E$ telle que l'on ait : $x' \circ \vec{m} = (x' \circ \vec{f}) . \mu$, pour tout $x' \in E'$.

Au couple $(\vec{m},\mu)$ on associe la probabilité cylindrique λ sur E définie de la manière suivante. Soit $C = \{x \in E ; (x'_i(x)) \in B \subset R^n, 1 \leqslant i \leqslant n\}$ un cylindre de base B (borélien de R^n) ; on pose :

$$\lambda(C) = \mu\{\omega \in \Omega ; (f_{x'_i}(\omega))_{1 \leqslant i \leqslant n} \in B\}$$

où $f_{x'_i}$ est évidemment la densité de Radon-Nikodym associée au couple $(x'_i \circ \vec{m}, \mu)$. Il est facile de voir que le nombre $\lambda(C)$ ne dépend pas du représentant choisi dans la classe $[f_{x'_i}]$.

Notons que s'il existe une densité faible $\vec{f} : \Omega \to E$ associée au couple $(\vec{m},\mu)$ alors (en vertu d'un résultat d'EDGAR (4) caractérisant la tribu de Baire d'un elc), la probabilité cylindrique λ coïncide avec la mesure image $\vec{f}(\mu)$, définie sur la tribu de Baire de E. Réciproquement, si λ est de Radon et si elle est concentrée à ε-près sur des disques compacts de E, alors il existe une densité faible $\vec{f} : \Omega \to E$ et on a évidemment $\lambda = \vec{f}(\mu)$ (pour plus de détails on pourra consulter (5) et (6)).

Rappelons qu'un espace complètement régulier T est dit universellement Radon-mesurable lorsqu'il est une partie universellement mesurable de son compactifié de Stone-Čech βT. L'espace T est dit encore radonien lorsque toute mesure borélienne positive et bornée sur T est de Radon (dans ce cas, T est bien sûr Radon universellement mesurable). On désignera enfin par $M_t^+(T)$ (resp. $M_\sigma^+(T)$) le cône positif des mesures de Radon (resp. des mesures de Baire) sur un espace complètement régulier T, cet espace de mesures étant muni de la topologie induite par la topologie étroite $\sigma(M(\beta T), C^\infty(T))$ de la convergence simple sur l'espace $C^\infty(T)$ des fonctions continues et bornées sur T.

1. PROPRIETE DE RADON-NIKODYM FAIBLE POUR LE CONE POSITIF $M_t^+(T)$ DES MESURES DE RADON SUR T.

Le résultat essentiel est résumé dans le théorème suivant :

1.1. THEOREME.

Pour tout couple $(\vec{m},\mu)$, *où* $\vec{m}$ *est une mesure à valeurs dans le cône positif* $M_t^+(T)$ *des mesures de Radon sur* T, *les conditions suivantes sont réalisées :*
a) il existe toujours une densité faible $\vec{f} : \Omega \to M_t^+(T)$ *associée au couple* $(\vec{m},\mu)$ *;*
b) la mesure image $\lambda = \vec{f}(\mu)$ *est de Radon sur* $M_t^+(T)$ *et de plus, pour tout* $\varepsilon > 0$ *elle est concentrée à* ε*-près sur des parties de* $M_t^+(T)$ *satisfaisant à la condition de PROKHOROV (1).*

Le point a) se démontre très facilement et fut d'ailleurs établi de manière indépendante par plusieurs auteurs (EDGAR, WEIZSACKER, etc...). Le point b) (dont la démonstration est un peu plus élaborée) mérite une attention plus particulière ; en effet, rien ne garantit a priori que la mesure image $\lambda = \vec{f}(\mu)$ soit de Radon et de plus, le fait que λ soit concentrée à ε-près sur des parties H_ε vérifiant la condition de Prokhorov est tout à fait exceptionnel, les compacts de $M_t^+(T)$ ne satisfaisant en général pas à cette propriété (sauf bien sûr si T est par exemple polonais ou localement compact). Notons à ce sujet que EDGAR, dans (3), démontre cette propriété dans un cas très particulier. Signalons encore que WEIZSACKER a établi que l'espace $M_t(T)$ des mesures de Radon (non nécessairement positives) sur T possède également la propriété de Radon-Nikodym faible ; j'ignore par contre si, dans ce cas, la mesure image $\lambda = \vec{f}(\mu)$ est encore de Radon.

Donnons, dans les grandes lignes, la démonstration du point b). On remarque tout d'abord que l'on peut écrire T sous la forme $(\bigcup_n K_n) \cup N$, où (K_n) est une suite de parties compactes de T deux à deux disjointes et N une partie $\vec{m}(\Omega)$-négligeable (donc $\vec{m}(A)$-négligeable, pour tout $A \in \Sigma$). Pour tout entier n, l'application $\vec{m}_n : \Omega \to M^+(K_n)$ définie par $\vec{m}_n(A) = 1_{K_n}.\vec{m}(A)$ est une mesure vectorielle, donc il existe une densité faible $\vec{f}_n : \Omega \to M^+(K_n)$ associée au couple $(\vec{m}_n, \mu)$. On vérifie ensuite que la série $\sum_n \vec{f}_n(\omega)$ converge dans $M^+(T)$ μ-presque partout, et que la fonction $\vec{f}$ définie par $\vec{f}(\omega) = \sum_n \vec{f}_n(\omega)$ aux points ω où la suite $(\vec{f}_n(\omega))$ est sommable et par $\vec{f}(\omega) = \vec{0}$ ailleurs est une densité faible associée au couple $(\vec{m},\mu)$. Il résulte alors du théorème d'Egoroff, que pour tout $\varepsilon > 0$, il existe $A \in \Sigma$, $\mu(T \setminus A) \leq \varepsilon$ de sorte que la partie $L_\varepsilon = \vec{f}(A) = \{\sum_n \vec{f}_n(\omega) \ ; \ \omega \in A\}$ satisfasse à la condition de Prokhorov. Soit $H_\varepsilon = \overline{\Gamma}_+(L_\varepsilon)$, l'enveloppe disquée fermée dans $M_t^+(T)$ de L_ε ; H_ε est une partie de $M_t^+(T)$ vérifiant encore la condition de Prokhorov et la mesure cylindrique λ est concentrée à ε-près sur H_ε. Les parties de Prokhorov étant relativement compactes dans $M_t(T)$, la mesure cylindrique λ est a fortiori de Radon sur $M_t^+(T)$, ce qui termine la preuve de b).

2. PROPRIETE DE RADON-NIKODYM FAIBLE POUR LE CONE POSITIF $M_\sigma^+(T)$ DES MESURES DE BAIRE SUR T.

Rappelons que l'on a toujours $M_\sigma^+(T) = M_\sigma^+(\nu T)$ où νT désigne le replété de Hewitt de T. Ainsi, on ne peut espérer obtenir des informations sur T à partir des propriétés de l'espace $M_\sigma^+(T)$ que si on a l'égalité $T = \nu T$ (T est dit alors "realcompact"). Dans toute la suite, nous supposerons donc que cette situation est toujours réalisée.

Le lien entre la propriété de Radon-Nikodym faible et la mesurabilité universelle de T dans βT se fait par l'intermédiaire du résultat suivant :

2.1. THEOREME.

Supposons que pour tout couple $(\vec{m},\mu)$, où $\vec{m}$ est une mesure à valeurs dans $M_\sigma^+(T)$, la mesure cylindrique associée λ soit de Radon. Alors l'espace T est Radon universellement mesurable.

PREUVE.

Pour le détail de la démonstration, on pourra consulter (5).

REMARQUE.

Lorsque la mesure λ associée au couple $(\vec{m},\mu)$ est de Radon sur $M_\sigma(T)$, elle est en fait portée par $M_\sigma^+(T)$ et il existe une densité faible associée $\vec{f} : \Omega \to M_\sigma(T)$; ceci résulte du fait que la classe des compacts de $M_\sigma(T)$ est stable par passage à l'enveloppe disquée fermée. J'ignore par contre si l'existence d'une densité faible garantie à elle seule que T soit Radon universellement mesurable dans βT.

Du théorème 2.1. on déduit, pour un espace métrisable T, le résultat suivant :

2.2. THEOREME.

Supposons que T soit métrisable et considérons les assertions :

a) T est radonien ;

b) T est Radon universellement mesurable ;

c) $M_\sigma(T) = M_t(T)$ *;*

d) Pour tout couple $(\vec{m},\mu)$*, avec* $\vec{m} : \Sigma \to M_\sigma^+(T)$*, la mesure cylindrique associée* λ *est de Radon ;*

e) Pour tout couple $(\vec{m},\mu)$*, avec* $\vec{m} : \Sigma \to M_\sigma^+(T)$*, il existe une densité faible* $\vec{f} : \Omega \to M_\sigma(T)$.

On a alors : a) $\Longleftrightarrow$ *b)* $\Longleftrightarrow$ *c)* $\Longleftrightarrow$ *d)* $\Rightarrow$ *e).*

PREUVE.

Le seul point à établir est l'équivalence entre les assertions a) et b). Celle-ci est évidente si on suppose que l'espace T est séparable et elle est démontrée par SCHACHERMAYER dans (9) moyennant des hypothèses de cardinalité sur T ; elle a finalement été établie par l'auteur dans le cas général dans (6). On pourra consulter (2) pour une démonstration détaillée.

J'ignore si, dans le cas général, la propriété e) équivaut aux autres assertions ; cela est par exemple le cas si l'espace métrisable T est séparable. En clair, on a la proposition suivante :

2.3. THEOREME.

Soit T un espace métrisable séparable. Alors toutes les assertions du théorème 2.2 sont équivalentes.

PREUVE.

Il suffit d'établir l'implication e) $\Longrightarrow$ b). Supposons que T ne soit pas Radon universellement mesurable ; alors, en vertu du théorème 2.1. il existe une probabilité $\mu \in M^+(\beta T)$ telle que l'on ait $\mu^*(T) = 1$ et $\mu_*(T) = 0$. On définit une mesure borélienne μ_T sur T par $\mu_T(B) = \mu(B')$ où B' est un borélien quelconque de βT vérifiant $B' \cap T = B$. On considère maintenant l'application mesurable $h : (T, \mathcal{B}(T)) \to (T \times \beta T, \mathcal{B}(T) \otimes \mathcal{B}(\beta T))$ définie par $h(t) = (t,t)$ et on note par ν la mesure image par l'application h de la mesure μ_T. Tout repose alors sur le lemme suivant :

LEMME.

Il n'existe pas de désintégration stricte (ν_s), $s \in \beta T$, *de la mesure* ν.

PREUVE DU LEMME.

Supposons qu'il existe une telle désintégration de ν. On vérifie alors aisément que pour tout $A \in \mathcal{B}(T) \otimes \mathcal{B}(\beta T)$, l'application $s \to \nu(A(.,s))$ est μ-mesurable et que l'on a $\nu(A) = \int \nu_s(A(.,s))d\mu(s)$. Or la diagonale $\Delta = \{(t,t) \; ; \; t \in T\}$ appartient à la tribu produit $\mathcal{B}(T) \otimes \mathcal{B}(\beta T)$ (qui, T étant métrisable séparable, coïncide avec la tribu borélienne de l'espace produit $T \times \beta T$) et l'on a $\nu(\Delta) = 1$; mais comme l'ensemble $B = \{s \ ; \ \nu_s(s) \neq 0\}$ est μ-négligeable, on a aussi $\int \nu_s(\Delta(.,s))d\mu(s) = 0$, d'où la contradiction.

Revenons à la preuve de e) $\Longrightarrow$ b). Considérons la mesure vectorielle $\vec{m} : \mathcal{B}(\beta T) \to M_\sigma^+(T)$ définie par $\vec{m}(A)(B) = \nu(A \times B) = \mu^*(A \cap B)$ pour tout $A \in \mathcal{B}(\beta T)$ et toute partie de Baire $B \in \mathcal{B}a(T)$. Par hypothèse, il existe une densité faible $\vec{f} : \beta T \to M_\sigma(T)$ associée au couple $(\vec{m},\mu)$. On vérifie aisément que la famille $(\vec{f}(s))$, $s \in \beta T$, est une désintégration de ν, ce qui est absurde en vertu du lemme précédent.

REMARQUE.

Lorsque T est métrisable séparable, on a évidemment $M_\tau(T) = M^\infty(T) = M_\sigma(T)$ (voir (1) et (6)). Il ne semble pas possible dans la démonstration précédente de se libérer de la condition "T est séparable". En effet, l'égalité des tribus $\mathcal{B}(T) \otimes \mathcal{B}(\beta T) = \mathcal{B}(T \times \beta T)$ est essentielle pour la démonstration du lemme et lorsque T est seulement métrisable, on n'a plus cette égalité. Tout ce que l'on peut dire se résume dans la proposition suivante :

2.4. PROPOSITION.

Soit (T,d) *un espace métrisable qui n'est pas Radon universellement-mesurable. Il existe alors un sous-espace fermé séparable* $X \subset T$ *tel que* $M_\sigma(X)$ *ne soit pas un espace de Radon-Nikodym faible.*

REMARQUE.

Notons que J.K. PACHL (8) a obtenu, par une méthode semblable à celle développée dans la preuve du lemme suivant le théorème 2.3., des résultats plus généraux sur la désintégration des mesures définies sur un produit d'espaces mesurés abstraits (article communiqué à l'auteur au cours de la Conférence).

3. DEUX EXEMPLES :

Pour terminer, donnons deux exemples, l'un concernant la désintégration, l'autre concernant la relation entre la dentabilité dans un elc et la propriété de Radon-Nikodym faible. Soit Ω un sous-espace de $[0,1]$ de mesure extérieure égale à 1 et de mesure intérieure nulle pour la mesure de Lebesgue μ. (Ω est évidemment non mesurable.) On note par η la mesure image sur l'espace produit $X = \Omega \times [0,1]$ par l'application $h : x \to (x,x)$.

EXEMPLE 1.

La mesure η est une mesure de Baire sur X n'admettant pas de désintégration

$(\eta_y)_{y \in [0,1]}$ constituée de mesures de Baire sur Ω. Ainsi, pour que toute mesure de Baire sur un espace produit T×Y soit désintégrable suivant $P_Y(\eta)$ il faut et il suffit (en vertu de (3) ou (7)) que sa projection sur T soit une mesure de Radon.

EXEMPLE 2.

Considérons l'espace $M_\sigma(\Omega)$ muni de la topologie initiale τ_i associée aux applications $\phi_d : M_\sigma(\Omega) \to \ell^1(N)$ définies par $\phi_d(\mu) = \sum_n \mu(f_n)$, où $d = (f_n)_n$ est une partition continue de l'unité sur Ω. Alors $M_\sigma(\Omega)$ est un espace complet dont toutes les parties bornées sont dentables et qui ne possède pas la propriété de Radon-Nikodym faible (5).

PREUVE.

Pour un espace complètement régulier T quelconque l'espace $M_\sigma(T)$ est complet (6) et toutes ses parties bornées sont dentables (5). Il résulte par ailleurs du théorème (2.3), que pour un tel espace Ω, l'espace $M_\sigma(\Omega)$ n'a pas la propriété de Radon-Nikodym faible.

BIBLIOGRAPHIE.

(1) A. BADRIKIAN, *Séminaire sur les fonctions aléatoires linéaires et les mesures cylindriques*, Lecture Notes, n° 139, 1970.

(2) D. BUCCHIONI et A. GOLDMAN, *Sur la convergence presque-partout des suites de fonctions mesurables*, Conference on vector space measures and applications, Dublin, 1977.

(3) G.A. EDGAR, *Disintegration of measures and the vector-valued Radon-Nikodym theorem*, Duke Math. Journal, vol. 42, n° 3, 1975, p. 447-450.

(4) G.A. EDGAR, *Mesurability in a Banach space*, Indiana Univ. Math. Journal, à paraître.

(5) A. GOLDMAN, *Mesures cylindriques, mesures vectorielles et questions de concentration cylindrique*, Pacific Journal of Math., vol. 69, n° 2, 1977, p. 40-69.

(6) A. GOLDMAN, Thèse d'Etat (en préparation).

(7) J. HOFFMAN - JØRGENSEN, *Existence of conditionnal probabilities*, Math. Scand. 28, 1971, p. 257-264.

(8) J.K. PACHL, *Disintégration and compact measures*, Math. Scand. (à paraître).

(9) W. SCHACHERMAYER, *Eberlein compacts et espaces de Radon*, C.R. Acad. Sc. Paris, 284, 1977, p. 405-407.

STABILITY OF TENSOR PRODUCTS OF RADON MEASURES OF TYPE ($\mathcal{H}$)

P. J. GUERRA
Facultad de C. Matemáticas
Univ. Complutense. Madrid-3
Spain

In [4] we studied simple convergence, simple convergence on every $H \in \mathcal{H}$, stability of the s-compactness and the s($\mathcal{H}$)-compactness of the image measure. In the present work we study this stability for the product by functions and the tensor product of measures.

Radon measures of type ($\mathcal{H}$) on an arbitrary topological space are studied in [6].

We will denote by E a topological space, Hausdorff or not, and by $\mathcal{G}, \mathcal{F}, \mathcal{K}$ and $\mathcal{B}$ the classes of subsets of E which are respectibily open, closed, closed-compact and Borel. $\mathcal{H}$ denotes a directed ($\subset$) class of closed subsets of E and M(E;$\mathcal{H}$) the set of Radon measures of type ($\mathcal{H}$) on E.

1. Preliminary definitions.

1. Definition. Let τ be a real set function, defined on $\mathcal{P}(E)$ (the set of all subsets of E), which verifies $\tau(\emptyset)=0$ and $\tau(X) \leq \tau(Y)$ when $X \subset Y$ ($\subset E$). A set $X \subset E$ is said to be τ-compact if, for every open cover $\mathcal{G}_0$ of X and $\varepsilon > 0$, there is a finite number of sets $G_k \in \mathcal{G}_0$ which satisfy the inequality

$$\tau(X - \bigcup_{k=1}^{n} G_k) < \varepsilon .$$

2. Definition. A Radon measure of type ($\mathcal{H}$) μ on E is a measure defined on the class $\mathcal{B}$ which satisfies the following two properties:

2.1. Every $H \in \mathcal{H}$ is μ-compact and $\mu(H) < +\infty$.

2.2. $\mu(B) = \sup\{\mu(H) \; : \; B \supset H \in \mathcal{H}\}$ for every $B \in \mathcal{B}$.

For every $X \in \mathcal{P}(E)$ and $H \in \mathcal{H}$ we define

$$\mu_{\bullet}(X) = \sup\{\mu(H) \; : \; X \supset H \in \mathcal{H}\},$$

$$\mu_H^*(X) = \mu(H) - \mu_\bullet(H-X)$$

and

$$\mu^\bullet(X) = \sup\{\mu_H^*(X \cap H) : H \in \mathcal{H}\}.$$

3. <u>Definition</u>. Let $(\mu_i)_{i\in I} \subset M(E;\mathcal{H})$, then $X \subset E$ is said to be $(\mu_i)_{i\in I}$-<u>compact</u> if, for every open cover $\mathcal{G}_\circ$ of X and $\varepsilon > 0$, there exists a finite number of open sets $G_k \in \mathcal{G}_\circ$ (k=1,...,n) such that

$$\overline{\lim_i}\, \mu_i^\bullet (X - \bigcup_{k=1}^n G_k) < \varepsilon$$

holds.

4. <u>Definition</u>. A net $(\mu_i)_{i\in I} \subset M(E;\mathcal{H})$ is <u>simply convergent</u> or s-<u>convergent</u> to $\mu \in M(E;\mathcal{H})$, denoted by $\mu_i \overset{s}{\dashrightarrow} \mu$, if and only if $\mu_i(B) \dashrightarrow \mu(B)$ for every $B \in \mathcal{B}$.

A net $(\mu_i)_{i\in I} \subset M(E;\mathcal{H})$ is <u>simply convergent on every</u> $H \in \mathcal{H}$ or $s(\mathcal{H})$-<u>convergent</u> to a measure $\mu \in M(E;\mathcal{H})$, and we write $\mu_i \overset{s(\mathcal{H})}{\dashrightarrow} \mu$, when $\mu_{iH} \overset{s}{\dashrightarrow} \mu_H$, μ_H and μ_{iH} being the Radon measures of type $\mathcal{H}' = \{H' \in \mathcal{H} : H' \subset H\}$ on H, induced by μ and μ_i respectively, for all $H \in \mathcal{H}$.

5. <u>Definition</u>. A net $(\mu_i)_{i\in I} \subset M(E;\mathcal{H})$ is said to be <u>simply compact</u> or s-<u>compact</u> (resp. $s(\mathcal{H})$-<u>compact</u>) if and only if every subnet has a s-convergent subnet (resp. $s(\mathcal{H})$-convergent).

2. <u>Stability of</u> $s(\mathcal{H})$-<u>compactness and</u> s-<u>compactness for the product by functions</u>.

Let μ be a Radon measure of type $(\mathcal{H})$ on E and p a non negative function μ-integrable on every $H \in \mathcal{H}$. Then by theorem 100 of [6], the real set function ν, defined by

$$\nu(B) = \sup\{\int_H p\, d\mu : B \supset H \in \mathcal{H}\} \quad (B \in \mathcal{B}),$$

belongs to $M(E;\mathcal{H})$ and furthemore we have $\nu(B) = \int_B p\, d\mu$ $(< +\infty)$ if $\mu(B) < +\infty$. This measure ν is called <u>product of</u> μ <u>by</u> p and is denoted by $p\mu$.

6. Theorem. If $(\mu_i)_{i\in I}$ is a net in $M(E;\mathcal{H})$ such that $\mu_i \xrightarrow{s(\mathcal{H})} \mu$ and p is a non negative bounded function on every $H\in\mathcal{H}$, which is integrable on every $H\in\mathcal{H}$ with respect to μ and μ_i for all $i\in I$. Then $p\mu_i \xrightarrow{s(\mathcal{H})} p\mu$ (1).

Proof. For every $B\in\mathcal{B}$, $H\in\mathcal{H}$ and $\varepsilon>0$ there exists a non negative function $f=\sum_{k=1}^{m}\alpha_k \chi_{A_k}$ with $\alpha_k\geq 0$ $(k=1,\dots,m)$ such that $f\leq p$ and

$$\begin{aligned}
p\mu(B\cap H) &= \int_{B\cap H} p\, d\mu \\
&\leq \sum_{k=1}^{m}\alpha_k\, \mu^{\bullet}(A_k\cap H\cap B)+\varepsilon \\
&= \lim_i \sum_{k=1}^{m}\alpha_k\, \mu_i^{\bullet}(A_k\cap H\cap B)+\varepsilon \\
&\leq \underline{\lim}_i \int_{B\cap H} p\, d\mu_i + \varepsilon \\
&= \underline{\lim}_i\, p\mu_i(B\cap H)+\varepsilon ,
\end{aligned}$$

from which we deduce the inequality

$$p\mu(B\cap H)\leq \underline{\lim}_i\, p\mu_i(B\cap H).$$

In a similar way, we can find a function $g=\sum_{k=1}^{n}\beta_k \chi_{A'_k}$ with $\beta_k=0$ $(k=1,\dots,n)$ verifying $p\leq g$ and

$$\begin{aligned}
\overline{\lim}_i\, p\mu_i(B\cap H) &= \overline{\lim}_i \int_{B\cap H} p\, d\mu_i \\
&\leq \overline{\lim}_i \sum_{k=1}^{n}\beta_k\, \mu_i(A'_k\cap B\cap H) \\
&= \sum_{k=1}^{n}\beta_k\, \mu(A'_k\cap B\cap H) \\
&\leq \int_{B\cap H} p\, d\mu + \varepsilon \\
&= p\mu(B\cap H)+\varepsilon .
\end{aligned}$$

Therefore, we have $p\mu(B\cap H) = \underline{\lim}_i\, p\mu_i(B\cap H) = \overline{\lim}_i\, p\mu_i(B\cap H)$ and

(1) We can find simple examples which show that this theorem is not true if we do not have p bounded on every $H\in\mathcal{H}$.

$p\mu_i \xrightarrow{s(\mathcal{H})} p\mu$.

7. Corollary. Assuming the conditions and notation of theorem 6 and if $\mu_i \xrightarrow{s} \mu$, then the following statements are equivalent:

7.1. $p\mu_i \xrightarrow{s} p\mu$.

7.2. For every Borel set $B \in \mathcal{B}$ with measure $p\mu(B) < +\infty$, and $\varepsilon > 0$ there exists $H \in \mathcal{H}$ such that $H \subset B$ and $\overline{\lim_i}\, p\mu_i(B-H) < \varepsilon$.

Proof. Indeed, if 7.1 is verified, then it follows from proposition 8 of [4] that 7.2 holds.

On the other hand, if 7.2 is assumed then 7.1 is easily deduced from theorem 6 and proposition 8 of [4].

8. Corollary. With the conditions and notation of theorem 6, if $\mu_i \xrightarrow{s} \mu$, p is bounded on E and $\mu(E) < +\infty$, then $p\mu_i \xrightarrow{s} p\mu$.

Proof. For every $B \in \mathcal{B}$ and $\varepsilon > 0$, it follows from theorem 6 and proposition 8 of [4], that there exists $H \in \mathcal{H}$ and $i_o \in I$ such that $H \subset B$, $\overline{\lim_i}\, \mu_i(B-H) < \varepsilon$ and

$$\begin{aligned}
\overline{\lim_i}\, p\mu_i(B-H) &\leq \sup_{i \geq i_o} p\mu_i(B-H) \\
&= \sup_{i \geq i_o} \int_{B-H} p \, d\mu_i \\
&\leq \sup_{x \in B} p(x) \sup_{i \geq i_o} \int_{B-H} d\mu_i \\
&= \sup_{x \in B} p(x) \sup_{i \geq i_o} \mu_i(B-H) \\
&\leq \sup_{x \in B} p(x) \cdot \varepsilon ,
\end{aligned}$$

Therefore, it follows from corollary 7 that $p\mu_i \xrightarrow{s} p\mu$.

9. Proposition. If every $H \in \mathcal{H}$ is regular with the induced topology, $H \cap F \in \mathcal{H}$ for all $H \in \mathcal{H}$, $F \in \mathcal{F}$, $(\mu_i)_{i \in I}$ is $s(\mathcal{H})$-compact in $M(E;\mathcal{H})$ and p is a non negative bounded function on every $H \in \mathcal{H}$ which is μ-integrable and μ_i-integrable on every $H \in \mathcal{H}$, for all $i \in I$, then $(p\mu_i)_{i \in I}$ is $s(\mathcal{H})$-compact.

Proof. For every $H \in \mathcal{H}$ it follows from theorem 10 of [4] that $\overline{\lim_i} \mu_i(H) < +\infty$, and so

$$\overline{\lim_i}\, p\mu_i(H) = \overline{\lim_i} \int_H p\, d\mu_i$$

$$\leq \sup_{x \in H} p(x)\ \overline{\lim_i} \int_H d\mu_i$$

$$= \sup_{x \in H} p(x)\ \overline{\lim_i}\, \mu_i(H)$$

$$< +\infty .$$

Moreover, for every $G \in \mathcal{G}$ and $H \in \mathcal{H}$ it follows from theorem 10 of [4] that for every open cover $\mathcal{G}_0$ of $G \cap H$ and $\varepsilon > 0$ there exists a finite number of open sets $G_1, \ldots, G_n \in \mathcal{G}_0$ such that

$$\overline{\lim_i}\, \mu_i(G \cap H - \bigcup_{k=1}^{n} G_k) < \varepsilon$$

and

$$\overline{\lim_i}\, p\mu_i(G \cap H - \bigcup_{k=1}^{n} G_k) = \overline{\lim_i} \int_{G \cap H - \bigcup_{k=1}^{n} G_k} p\, d\mu_i$$

$$\leq \sup_{x \in H} p(x)\ \overline{\lim_i}\, \mu_i(G \cap H - \bigcup_{k=1}^{n} G_k)$$

$$\leq \sup_{x \in H} p(x) \cdot \varepsilon \quad ,$$

therefore, it follows inmediately that $G \cap H$ is $(p\mu_i)_{i \in I}$-compact and so it follows from theorem 10 of [4] that $(p\mu_i)_{i \in I}$ is $s(\mathcal{H})$-compact.

10. Corollary. Assuming the conditions and notation of proposition 9 we have that if $\overline{\lim_i}\, \mu_i(E) < +\infty$, p is bounded on E and $(\mu_i)_{i \in I}$ is s-compact then the net $(p\mu_i)_{i \in I}$ is s-compact.

Proof. Obviously we can find $i_0 \in I$ such that $\sup_{i \geq i_0} p\mu_i(E) < +\infty$ and

$$\overline{\lim_i}\, p\mu_i(E) \leq \sup_{i \geq i_0} p\mu_i(E)$$

$$= \sup_{i \geq i_0} \int_E p\, d\mu_i$$

$$= \sup_{x \in E} p(x)\ \sup_{i \geq i_0} \mu_i(E)$$

$< +\infty .$

From theorem 10 of [4] we deduce that every $H \in \mathcal{H}$ is $(p\mu_i)_{i \in I}$-compact, because according to proposition 9 the net $(p\mu_i)_{i \in I}$ is $s(\mathcal{H})$-compact. Moreover, it follows from theorem 11 of [4] that for every Borel set $B \in \mathcal{B}$ and $\varepsilon > 0$, there exists $H \in \mathcal{H}$ verifying $H \subset B$ and $\overline{\lim_i}\, \mu_i(B-H) < \varepsilon$. Since $\overline{\lim_i}\, \mu_i(B) < +\infty$ we can find $i_1 \in I$ such that $\mu_i(B) < +\infty$ and $\mu_i(B-H) < \varepsilon$ for all $i \in I$ with $i \geq i_1$, therefore, we have

$$\begin{aligned}\overline{\lim_i}\, p\mu_i(B-H) &\leq \sup_{i \geq i_1} p\mu_i(B-H)\\ &= \sup_{i \geq i_1} \int_{B-H} p \, d\mu_i\\ &\leq \sup_{x \in E} p(x) \sup_{i \geq i_1} \mu_i(B)\\ &\leq \varepsilon \cdot \sup_{x \in E} p(x)\end{aligned}$$

and so we easily deduce from theorem 11 of [4] that the net $(p\mu_i)_{i \in I}$ is s-compact.

3. Stability of $s(\mathcal{H})$-compactness for tensor products .

Let E_1, E_2 be two topological spaces and $\mu^k \in M(E_k; \mathcal{H}_k)$ $(k=1,2)$, where each $H_k \in \mathcal{H}_k$ is a regular topological space for the induced topology. Then, if we denotes by $\mathcal{H}$ the class of all closed subsets of $E = E_1 x E_2$ contained in some $H_1 x H_2$ $(H_k \in \mathcal{H}_k, k=1,2)$, it follows from theorem 103 of [6] that there exists one and only one measure $\mu \in M(E;\mathcal{H})$ which verifies $\mu(B_1 x B_2) = \mu^1(B_1) \cdot \mu^2(B_2)$ for each pair (B_1, B_2) of Borel sets of E_1 and E_2 ([3]). This measure is called the tensor product in $M(E;\mathcal{H})$ of μ^1 and μ^2, and is denoted by $\mu^1 \otimes \mu^2$.

11. Theorem. Assuming the conditions and notation of the last paragraph and let $(\mu_i^k)_{i \in I}$ denote a $s(\mathcal{H}_k)$-compact net in $M(E_k; \mathcal{H}_k)$ for $k=1,2$, then the following assertions are equivalent:

11.1 the net $(\mu_i^1 \otimes \mu_i^2)_{i \in I}$ is $s(\mathcal{H})$-compact,

11.2 for every open set $G \in \mathcal{G}$, $H_1 x H_2 \in \mathcal{H}$ and $\varepsilon > 0$ there exists a

([3]) We take here $0 \cdot \infty = 0$.

finite number of open sets $G_k'xG_k'' \in \mathcal{G}$ (k=1,...,n) such that

$$(\bigcup_{k=1}^{n} G_k'xG_k'') \cap (H_1xH_2) \subset G \cap (H_1xH_2)$$

and

$$\overline{\lim_i} \mu_i^1 \otimes \mu_i^2 (G \cap H_1xH_2 - \bigcup_{k=1}^{n} G_k'xG_k'') < \varepsilon .$$

Proof. 11.1 implies 11.2: If $\mu_i = \mu_i^1 \otimes \mu_i^2$ for every $i \in I$, it follows from theorem 10 of [4] that $G \cap (H_1xH_2)$ is $(\mu_i)_{i\in I}$-compact for all $G \in \mathcal{G}$ and $H_1xH_2 \in \mathcal{H}$, from which immediately deduce 11.2, since the open sets of the form G'xG" are a base for the product topology.

11.2 implies 11.1: It follows from theorem 10 of [4] that $\overline{\lim_i}\mu_i^k(H_k) < +\infty$ for every $H_k \in \mathcal{H}_k$ (k=1,2), from which it is easily deduced that $\overline{\lim_i} \mu_i (H_1xH_2) < +\infty$. Moreover any $H \in \mathcal{H}$ is contained, by definition, in some $H_1xH_2 \in \mathcal{H}$, and so we have that $\overline{\lim_i} \mu_i(H) < +\infty$ holds for all $H \in \mathcal{H}$.

To prove that $G \cap H$ is $(\mu_i)_{i\in I}$-compact for all $H \in \mathcal{H}$ and $G \in \mathcal{G}$, it is enough to show it in the particular case of $H=H_1xH_2$, because every $H \in \mathcal{H}$ is contained in some set of this form, and we know according to definition 3 of [4], that every closed subset of a $(\mu_i)_{i\in I}$-compact set is also $(\mu_i)_{i\in I}$ -compact. Moreover, to prove that $G \cap H_1xH_2$, where $G \in \mathcal{G}$ and $H_1xH_2 \in \mathcal{H}$, is $(\mu_i)_{i\in I}$-compact, it is enough to do it in tha particular case of $G=G_1xG_2$, with $G_i \in \mathcal{G}_i$ (i=1,2). So, let us show that $(G_1xG_2) \cap (H_1xH_2)$ is $(\mu_i)_{i\in I}$-compact, for all $G_k \in \mathcal{G}_k$ and $H_k \in \mathcal{H}_k$ (k=1,2). Let $(G_\lambda)_{\lambda\in\Lambda}$ be an open cover of $(G_1xG_2) \cap (H_1xH_2)$ (we can of course suppose that $G_\lambda = G_\lambda' \; xG_\lambda''$, where G_λ', G_λ'' are open sets of E_1 and E_2 respectively). For every $x \in G_1 \cap H_1$, let Λ_x be the set of all $\lambda \in \Lambda$ such that $x \in G_\lambda' \cap H_1$. Then, $G_2 \cap H_2 \subset \bigcup_{\lambda\in\Lambda_x} G_\lambda''$ and since $(\mu_i^k)_{i\in I}$ is $s(\mathcal{H}_k)$-compact, it follows from theorem 10 of [4] that $G_k \cap H_k$ is $(\mu_i^k)_{i\in I}$-compact for k=1,2, and therefore, there exists for every $\varepsilon > 0$, a finite subset $\Lambda_x' \subset \Lambda_x$ for which

$$\overline{\lim_i} \mu_i^2 (G_2 \cap H_2 - \bigcup_{\lambda\in\Lambda_x'} G_\lambda'') < \varepsilon$$

holds.

Let $V_x = \bigcap_{\lambda\in\Lambda_x} G_\lambda'$, then $x \in V_x$ and $\{V_x\}_{x\in G_1\cap H_1}$ is an open cover of $G_1 \cap H_1$, so we can find a finite number of elements $x_1,\ldots,x_n \in G_1 \cap H_1$ such that

$$\overline{\lim_i}\,\mu_i^1(G_1 \cap H_1 - \bigcup_{k=1}^{n} V_{x_k}) < \varepsilon$$

holds.

Then, if $\Lambda_k' = \Lambda_{x_k}'$, $V_k = V_{x_k}$ and $V_k' = V_k - \bigcup_{j=1}^{k-1} V_{x_j}$ $(k=1,\dots,n)$, the following hold:

$$(G_1 \cap H_1) \times (G_2 \cap H_2) - \bigcup_{k=1}^{n} \bigcup_{\lambda \in \Lambda_k'} G_\lambda' \times G_\lambda'' \subset \Big[\bigcup_{k=1}^{n} (V_k' \cap G_1 \cap H_1) \times ((G_2 \cap H_2) - \bigcup_{\lambda \in \Lambda_k'} G'') \cup$$

$$\cup ((G_1 \cap H_1) - \bigcup_{k=1}^{n} V_k) \times (G_2 \cap H_2)\Big]$$

and

$$\overline{\lim_i}\,\mu_i \Big[(G_1 \cap H_1) \times (G_2 \cap H_2) - \bigcup_{k=1}^{n} \bigcup_{\lambda \in \Lambda_k'} G_\lambda' \times G_\lambda''\Big] \leq \overline{\lim_i}\,\mu_i \Big[\bigcup_{k=1}^{n} (V_k' \cap G_1 \cap H_1) \times$$

$$\times ((G_2 \cap H_2) - \bigcup_{\lambda \in \Lambda_k'} G_\lambda'') + \varepsilon \cdot \overline{\lim_i}\,\mu_i (G_2 \cap H_2).$$

Since the inequality

$$\overline{\lim_i}\,\mu_i^2 \Big[(G_2 \cap H_2) - \bigcup_{\lambda \in \Lambda_k'} G_\lambda''\Big] < \varepsilon$$

holds for $k=1,\dots,n$ we can find $i_k \in I$ $(k=1,\dots,n)$ such that

$$\mu_i^2 \Big[(G_2 \cap H_2) - \bigcup_{\lambda \in \Lambda_k'} G_\lambda''\Big] < \varepsilon$$

is verified for all $i \geq i_k$ $(i \in I)$.

Let $i_\circ \in I$ be such that $i_\circ \geq i_k$ for $k=1,\dots,n$, then we have

$$\overline{\lim_i}\,\mu_i \Big[\bigcup_{k=1}^{n} (V_k' \cap G_1 \cap H_1) \times ((G_2 \cap H_2) - \bigcup_{\lambda \in \Lambda_k'} G_\lambda'')\Big] = \overline{\lim_i} \sum_{k=1}^{n} \mu_i^1 (V_k' \cap G_1 \cap H_1) \mu_i^2 \Big[(G_2 \cap H_2) -$$

$$- \bigcup_{\lambda \in \Lambda_k'} G'\Big]$$

$$\leq \inf_{i \geq i_\circ} \sup_{j \geq i} \sum_{k=1}^{n} \mu_j^1 (V_k' \cap G_1 \cap H_1)\, \mu_j^2 \Big[(G_2 \cap H_2) -$$

$$- \bigcup_{\lambda \in \Lambda_k'} G'\Big]$$

$$\leq \varepsilon \inf_{i \geq i_\circ} \sup_{j \geq i} \sum_{k=1}^{n} \mu_j^1 (V_k' \cap G_1 \cap H_1)$$

$$= \varepsilon\, \overline{\lim_i} \sum_{k=1}^{n} \mu_i^1 (\bigcup_{k=1}^{n} V_k' \cap G_1 \cap H_1)$$

$$\leq \varepsilon \ \overline{\lim_i} \mu_i^1(G_1 \cap H_1).$$

Therefore,

$$\overline{\lim_i} \mu_i \left[(G_1 \cap H_1) x (G_2 \cap H_2) - \bigcup_{k=1}^{n} \bigcup_{\lambda \in \Lambda_k'} G_\lambda' x G_\lambda''\right] \leq \varepsilon \left[\overline{\lim_i} \mu_i^1(G_1 \cap H_1) + \overline{\lim_i} \mu_i^2(G_2 \cap H_2)\right]$$

holds and $(G_1 x G_2) \cap (H_1 x H_2)$ is $(\mu_i)_{i \in I}$-compact.

Now we deduce from theorem 10 of [4] that $(\mu_i)_{i \in I}$ is $s(\mathcal{H})$-compact, because $\overline{\lim_i} \mu_i(H) < +\infty$ for all $H \in \mathcal{H}$ and $G \cap H$ is $(\mu_i)_{i \in I}$-compact for every $G \in \mathcal{G}$ and $H \in \mathcal{H}$.

12.<u>Corollary</u>. With the notation of theorem 11 and if the net $(\mu_i^k)_{i \in I}$ is s-compact and $\overline{\lim_i} \mu_i^k(E_k) < +\infty$ for k=1,2, then the net $(\mu_i^1 \otimes \mu_i^2)_{i \in I}$ is s-compact if and only if 11.2 is verified.

<u>Proof</u>. Evidently, if the net $(\mu_i^1 \otimes \mu_i^2)_{i \in I}$ is s-compact then it is $s(\mathcal{H})$-compact and it follows from theorem 11 that 11.2 is verified.

Let us now suppose that 11.2 is satisfied. Then according to theorem 11 the net $(\mu_i^1 \otimes \mu_i^2)_{i \in I}$ is $s(\mathcal{H})$-compact and it follows from theorem 10 of [4] that every $H \in \mathcal{H}$ is $(\mu_i^1 \otimes \mu_i^2)_{i \in I}$ -compact. Moreover, for every $G \in \mathcal{G}$ and $\varepsilon > 0$ we can find $H_k \in \mathcal{H}_k$ such that $H_k \subset \pi_k(G) = G_k$ and $\overline{\lim_i} \mu_i^k(G_k - H_k) < \varepsilon'$ for k=1,2, $\pi_k : E \dashrightarrow E_k$ being the projection (k=1,2) and $\varepsilon' = \min\left(\sqrt{\frac{\varepsilon}{6}}, \frac{\varepsilon}{6 \overline{\lim_i} \mu_i^2(G_2)}, \frac{\varepsilon}{6 \overline{\lim_i} \mu_i^1(G_1)}\right)$ (4).

Then,

$$\overline{\lim_i} \mu_i^1 \otimes \mu_i^2 (G - H_1 x H_2) \leq \overline{\lim_i} \mu_i^1 \otimes \mu_i^2 (G_1 x G_2 - H_1 x H_2)$$

$$\leq \overline{\lim_i} \mu_i^1(G_1 - H_1) \ \overline{\lim_i} \mu_i^2(G_2 - H_2) + \overline{\lim_i} \mu_i^1 G_1) \cdot$$

$$\cdot \overline{\lim_i} \mu_i^2(G_2 - H_2) + \overline{\lim_i} \mu_i^1(G_1 - H_1) \ \overline{\lim_i} \mu_i^2(G_2)$$

$$< \frac{\varepsilon}{2} \quad .$$

---

(4) The case where $\overline{\lim_i} \mu_i^k(G_k) = 0$, is trivial (k=1,2).

If $G \cap H_1 x H_2 = \emptyset$, then

$$\overline{\lim_i} \mu_i^1 \otimes \mu_i^2 (G-H) \leq \overline{\lim_i} \mu_i^1 \otimes \mu_i^2 (G - H_1 x H_2)$$

$$\leq \overline{\lim_i} \mu_i^1 \otimes \mu_i^2 (G_1 x G_2 - H_1 x H_2)$$

$$\leq \varepsilon$$

holds for all $H \in \mathcal{H}$, and in the case where $G \cap H_1 x H_2 \neq \emptyset$, since $H_1 x H_2$ is $(\mu_i^1 \otimes \mu_i^2)_{i \in I}$-compact and regular for the induced topology, we can find $H \in \mathcal{H}$ such that $H \subset G \cap (H_1 x H_2) \subset G$ and

$$\overline{\lim_i} \mu_i^1 \otimes \mu_i^2 [G \cap (H_1 x H_2) - H] < \frac{\varepsilon}{2}$$

is satisfied. Therefore

$$\overline{\lim_i} \mu_i^1 \otimes \mu_i^2 (G-H) \leq \lim_i \mu_i^1 \otimes \mu_i^2 (G \cap (H_1 x H_2) - H) + \overline{\lim_i} \mu_i^1 \otimes \mu_i^2 (G - (H_1 x H_2)) \leq \varepsilon$$

holds.

So it follows from theorem 11 of [4] that the net $(\mu_i^1 \otimes \mu_i^2)_{i \in I}$ is s-comapcat since

$$\overline{\lim_i} \mu_i^1 \otimes \mu_i^2 (E) \leq \lim_i \mu_i^1 (E_1) \lim_i \mu_i^2 (E_2) < +\infty .$$

13. Proposition. With the notation of theorem 11, and if $(\mu_i^k)_{i \in I}$ is a net in $M(E; \mathcal{H}_k)$ such that $\mu_i^k \xrightarrow{s(\mathcal{H}_k)} \mu^k$ (k=1,2) then $\mu_i^1 \otimes \mu_i^2 \xrightarrow{s(\mathcal{H})} \mu^1 \otimes \mu^2$ if and only if 11.2 is satisfied.

Proof. If $\mu_i^1 \otimes \mu_i^2 \xrightarrow{s(\mathcal{H})} \mu^1 \otimes \mu^2$ then the net $(\mu_i^1 \otimes \mu_i^2)_{i \in I}$ is $s(\mathcal{H})$-compact and 11.2 follows immediately from theorem 11.

Let us suppose now that 11.2 holds and denote by μ_i the tensor product $\mu_i^1 \otimes \mu_i^2$ and $\mu^1 \otimes \mu^2$ by μ. Then it follows from theorem 11 that the net $(\mu_i)_{i \in I}$ is $s(\mathcal{H})$-compact, so every subnet $(\mu_j)_{j \in J}$ $(J \subset I)$ has a subnet $(\mu_{j'})_{j' \in J'}$ $(J' \subset J)$ which is $s(\mathcal{H})$-convergent to a measure $\mu_{J'} \in M(E; \mathcal{H})$. Since the net $(\mu_{j'}^k)_{j' \in J'}$ is $s(\mathcal{H}_k)$-convergent to μ^k, we have

$$\mu_{J'} [(B_1 \cap H_1) x (B_2 \cap H_2)] = \lim_{j'} \mu_{j'} [(B_1 \cap H_1) x (B_2 \cap H_2)]$$

$$= \lim_{j'} \mu^1_{j'}(B_1 \cap H_1) \lim_{j'} \mu^2_{j'}(B_2 \cap H_2)$$

$$= \mu^1(B_1 \cap H_1)\ \mu^2(B_2 \cap H_2)$$

for all $B_k \in \mathcal{B}_\kappa$ and $H_k \in \mathcal{H}_k$ (K=1,2), so it follows easily that $\mu_{J'} = \mu^1 \otimes \mu^2$. Then, as we have just proved, every subnet of the net $(\mu_i)_{i\in I}$ has a subnet $s(\mathcal{H})$-convergent to $\mu^1 \otimes \mu^2$ and therefore, the net $(\mu_i)_{i\in I}$ is also $s(\mathcal{H})$-convergent to $\mu^1 \otimes \mu^2$.

14. Corollary. With the notation of proposition 13 and if $\mu^k_i \xrightarrow{s} \mu^k$ and $\overline{\lim_i} \mu^k_i(E_i) < +\infty$ hold for k=1,2, then $\mu^1_i \otimes \mu^2_i \xrightarrow{s} \mu^1 \otimes \mu^2$ if and only if 11.2 is verified.

Proof. It is evident that when the net $(\mu^1_i \otimes \mu^2_i)_{i\in I}$ is s-convergent to $\mu^1 \otimes \mu^2$, then the net is $s(\mathcal{H})$-convergent to the tensor product of μ^1 and μ^2 and it follows from proposition 13 that 11.2 is verified.

If we assume that 11.2 holds, then according to corollary 12 the net $(\mu^1_i \otimes \mu^2_i)_{i\in I}$ is s-convergent to a measure $\mu_{J'} \in M(E;\mathcal{H})$. Furthermore, it follows from proposition 13 that $(\mu^1_{j'} \otimes \mu^2_{j'})_{j'\in J'}$ is $s(\mathcal{H})$-convergent to $\mu^1 \otimes \mu^2$, and since this limit is unique and s-convergence implies the $s(\mathcal{H})$-convergence, we deduce that $\mu_{J'} = \mu^1 \otimes \mu^2$ and the net $(\mu^1_i \otimes \mu^2_i)_{i\in I}$ is s-convergent to $\mu^1 \otimes \mu^2$.

References

[1] BILLINGSLEY, P.: Convergence of probability measures. John Wiley. New York. 1968.

[2] DIEUDONNE, J. : Sur la convergence des suites de mesures de Radon. Anais Acad. Brasil Ci. 23 (1951), 21-38, 277-282.

[3] GROTHENDIECK, A.: Sur les applications linéaires faiblement compactes d'espaces du type c(K). Canadian J. Math. 5 (1953), 129-173.

[4] JIMENEZ GUERRA, P.: Compactness in the space of Radon measures of type ($\mathcal{H}$). To appear in Proc. R. Irish Acad.

[5] PROKOROV, Yu. V.: Convergence of random processes and limit theorems in probability theory. Probab. Appl. 1 (1956), 157-214.

[6] RODRIGUEZ-SALINAS,B. y P. J. GUERRA: Medidas de Radon de tipo ($\mathcal{H}$) en espacios topológicos arbitrarios. To appear in Mem. R. Acad. Ci. Madrid.

[7] SCHWARTZ,L. : Radon measures on arbitrary topological spaces and cylindrical measures. Oxford University Press, 1973.

[8] TOPSØE, F. : Compactness in spaces of measures. Studia Math. XXXVI (1970), 194-212.

THE STRONG MARKOV PROPERTY FOR CANONICAL WIENER PROCESSES

R.L.Hudson
Mathematics department, Nottingham University, Nottingham, England.

1. Canonical Wiener Processes

By a canonical Wiener process of variance $\sigma^2 \geq 1$ indexed by the positive half-line $R_{\geq 0}$, we mean a triple (P,Q,ψ), where P,Q are functions from $R_{\geq 0}$ to the self-adjoint operators in a hilbert space H and Ψ is a unit vector in H, with the following properties.

0). $[P(s),P(t)] = [Q(s),Q(t)] = 0$, $[P(s),Q(t)] = -i \min\{s,t\}$.

1). $P(0) = Q(0) = 0.$

2). For disjoint intervals $\Delta_1=(a_1,b_1],\ldots,\Delta_n=(a_n,b_n]\subset R_{\geq 0}$, denoting by (p_Δ,q_Δ) the canonical pair $(\sqrt{(b-a)}(P(b)-P(a)),\sqrt{(b-a)}(Q(b)-Q(a)))$ when $\Delta=(a,b]$, the canonical pairs $(p_{\Delta_1},q_{\Delta_1}),\ldots,(p_{\Delta_n},q_{\Delta_n})$ are independent and identically normally distributed with zero means and covariance matrix $\frac{1}{2}\sigma^2 I_2$ in the state determined by Ψ.

Here the commutation relation $[X,Y]=-ic$, where c is a real number and X,Y are unbounded self-adjoint operators means the corresponding Weyl relation $e^{ixX}e^{iyY}=e^{icxy}e^{iyY}e^{ixX}$. A canonical pair is a pair of self-adjoint operators (p,q) satisfying $[p,q]=-i1$ and the definitions of independence, identity of distribution, normal distribution, mean and covariance matrix for canonical pairs are as in [3],[2].

Canonical Wiener processes indexed by the unit interval were introduced in [2] and shown to have some properties analogous to those of classical Wiener processes in [1],[2].

A canonical Wiener process (P_o,Q_o,Ψ_o) with $\sigma^2=1$ is obtained from the Fock representation of the canonical commutation relations over the real Hilbert space $\mathfrak{R}$ of square-integrable functions on $R_{\geq 0}$ by taking Ψ_o to be the Fock vacuum vector and setting $P_o(t)=\pi(\chi_{[0,t]})$, $Q_o(t)=\varphi(\chi_{[0,t]})$, where for $f\in\mathfrak{R}$ $\pi(f),\varphi(f)$ are the canonical field operators and $\chi_{[0,t]}$ is the indicator function of $[0,t]$. When $\sigma^2>1$, a canonical Wiener process (P,Q,Ψ) can be constructed in the Hilbert space tensor product $H=H_o\otimes H_o$ of two copies of Fock space by taking Ψ to be $\Psi_o\otimes\Psi_o$ and setting

$$P(t)=2^{-\frac{1}{2}}(\alpha P_o(t)\otimes 1+\alpha^{-1}1\otimes Q_o(t)),\quad Q(t)=2^{-\frac{1}{2}}(\alpha^{-1}Q_o(t)\otimes 1-\alpha 1\otimes P_o(t)),\qquad (1.1)$$

where α is a real number such that $\alpha^2+\alpha^{-2}=2\sigma^2$. These processes are cyclic, meaning that repeated action on the state vector by the constituent operators of the process yields a total set of vectors, and every cyclic process is unitarily equivalent to one of this type.

We denote by N the von Neumann algebra generated by the spectral projections of the constituent operators of the process (P,Q,Ψ), and for $\lambda \geq 0$ by N_λ (resp. ${}_\lambda N$) the <u>pre-</u> (resp. <u>post-</u>) <u>λ algebra</u>, generated by the spectral projections of $P(t), Q(t), t \leq \lambda$ (resp. of $P(t+\lambda)-P(\lambda)$, $Q(t+\lambda)-Q(\lambda), t \geq 0$). $N_\lambda, {}_\lambda N$ are <u>independent</u> in the sense that if $A \varepsilon N_\lambda$, $B \varepsilon {}_\lambda N$ then A, B commute and $\langle AB\Psi, \Psi\rangle = \langle A\Psi, \Psi\rangle\langle B\Psi, \Psi\rangle$.

A positive self-adjoint operator T with spectral resolution $T = \int_0^\infty \lambda dE(\lambda)$ is a <u>Markov time</u> if $E(\lambda) \varepsilon N_\lambda$ for all $\lambda \geq 0$.

2. <u>Existence of</u> P_T, Q_T

Analogy with the strong Markov property for a classical Wiener process [5] suggests that if $T = \int_0^\infty \lambda dE(\lambda)$ is a Markov time for the cyclic canonical Wiener process (P, Q, Ψ) then (P_T, Q_T, Ψ) is also a canonical Wiener process, where formally

$$P_T(t) = \int_0^\infty (P(t+\lambda)-P(\lambda))dE(\lambda), \quad Q_T(t) = \int_0^\infty (Q(t+\lambda)-Q(\lambda))dE(\lambda). \tag{2.1}$$

To give meaning to (2.1) we first write down the equivalent forms

$$e^{ixP_T(t)} = \int_0^\infty e^{ixP(t+\lambda)} e^{-ixP(\lambda)} dE(\lambda), \tag{2.2}$$

$$e^{ixQ_T(t)} = \int_0^\infty e^{ixQ(t+\lambda)} e^{-ixQ(\lambda)} dE(\lambda), \tag{2.3}$$

and observe that the integrands in (2.2),(2.3) belong to ${}_\lambda N$ whereas the integrator belongs to N_λ, suggesting that the integrals be defined as strong operator limits of 'backward' Riemann-Stieltjes sums

$$K(\Lambda) = \sum_{j=1}^{n} e^{ixP(t+\lambda_j)} e^{-ixP(\lambda_j)} (E(\lambda_j)-E(\lambda_{j-1})), \tag{2.4}$$

$$L(\Lambda) = \sum_{j=1}^{n} e^{ixQ(t+\lambda_j)} e^{-ixQ(\lambda_j)} (E(\lambda_j)-E(\lambda_{j-1})). \tag{2.5}$$

where $b>0$ and $\Lambda = \{0=\lambda_0<\lambda_1<\ldots<\lambda_n=b\}$ is a partition of $[0,b]$.

<u>Theorem 2.1</u> As $\mathrm{Max}(\lambda_j-\lambda_{j-1}) \to 0$ and $b \to \infty$ $K(\Lambda), L(\Lambda)$ converge strongly to operators $U_{x,t}, V_{x,t}$ respectively. Moreover for fixed t, $x \mapsto U_{x,t}$, $x \mapsto V_{x,t}$ are strongly continuous one-parameter unitary groups whose infinitesimal generators $P_T(t), Q_T(t)$ satisfy the defining properties 0),1) of a canonical Wiener process.

<u>Lemma 2.2</u> For fixed t, $(x,\lambda) \mapsto e^{ixP(t+\lambda)} e^{-ixP(\lambda)}$ is strongly continuous on $R \times R_{>0}$.

<u>Proof</u> By (1.1) in the case $\sigma^2>1$ these operator-valued functions are tensor products of corresponding functions for the Fock case $\sigma^2=1$. Hence it is sufficient to consider the latter case. Using the Fock vacuum expectation functional $\langle e^{i\pi(f)}\Psi, \Psi\rangle = e^{-\frac{1}{4}\|f\|^2}$, we have after some manipulations, for arbitrary $f, g \varepsilon \mathfrak{M}, (x,\lambda), (y,\mu) \varepsilon R \times R_{>0}$,

$$\|(e^{ixP_0(t+\lambda)} e^{-ixP_0(\lambda)} - e^{iyP_0(t+\mu)} e^{-iyP_0(\mu)}) e^{i\pi(f)} e^{i\varpi(g)}\Psi\|^2$$

$$= 2(1-\cos(\langle xI_{(\lambda,t+\lambda]} - yI_{(\mu,t+\mu]}, g\rangle) e^{-\frac{1}{4}\|xI_{(\lambda,t+\lambda]} - yI_{(\mu,t+\mu]}\|^2}.$$

From this it is clear that $(x,\lambda) \mapsto e^{ixP(t+\lambda)} e^{-ixP(\lambda)} e^{i\pi(f)} e^{i\varpi(g)}\Psi$ is continuous. Since Ψ is cyclic vectors of the form $e^{i\pi(f)} e^{i\varpi(g)}\psi$ are total in H_0 and the result follows.

<u>Lemma 2.3</u> There exists a total set S of which Ψ is an element such

that, for arbitrary $\chi_1, \chi_2 \in S$, $\langle \chi_1, \chi_2 \rangle \neq 0$ and

$$\langle AB\chi_1, \chi_2 \rangle = \langle \chi_1, \chi_2 \rangle^{-1} \langle A\chi_1, \chi_2 \rangle \langle B\chi_1, \chi_2 \rangle \tag{2.6}$$

whenever, for some $\lambda \geq 0$, $A \in N_\lambda$, $B \in {}_\lambda N$.

Proof In the Fock case $\sigma^2 = 1$, $S = S_o$ is the set of 'exponential vectors' which is wellknown to be total and includes the vacuum [4]. For each $\lambda \geq 0$ these vectors are product vectors for the tensor product decomposition $H_o = H_{o\lambda} \otimes {}_\lambda H_o$ of Fock space H_o into the Fock spaces $H_{o\lambda}$, ${}_\lambda H_o$ over the subspaces of $\Re$ whose elements vanish outside $[0,\lambda]$, $[\lambda,\infty)$ respectively. Since $N_\lambda = B(H_{o\lambda}) \otimes 1$, ${}_\lambda N = 1 \otimes B({}_\lambda H_o)$, (2.6) follows. In the case $\sigma^2 > 1$ we use the realisation (1.1) and take S to be the set $S_o \otimes S_o$ of all product vectors in $H = H_o \otimes H_o$ formed from elements of S_o. It is easy to see, using the corresponding property in the Fock case, that for each $\lambda \geq 0$ there is a tensor product decomposition $H = H_\lambda \otimes {}_\lambda H$ for which the elements of S are product vectors and such that $N_\lambda \subset B(H_\lambda) \otimes 1$, ${}_\lambda N \subset 1 \otimes B({}_\lambda H)$. Since $\psi = \psi_o \otimes \psi_o \in S$ and S is clearly total the result follows.

Lemma 2.4 For $\chi \in S$, $K(\Lambda)\chi$ converges.

Proof We write $F(\lambda) = e^{ixP(t+\lambda)} e^{-ixP(\lambda)}$. Then if Λ_1, Λ_2 are partitions of $[0,b]$ whose union is $\{0 = \lambda_o < \lambda_1 < \ldots < \lambda_n = b\}$ and if λ_k^1 (resp. λ_k^2) is the smallest element of Λ_1 (resp. Λ_2) not less than λ_k, using Lemma 2.3 we have

$$\begin{aligned}
\|(K(\Lambda_1) - K(\Lambda_2))\chi\|^2 &= \|\sum_k \{(F(\lambda_k^1) - F(\lambda_k^2))(E(\lambda_k) - E(\lambda_{k-1}))\chi\}\|^2 \\
&= \|\sum_k \{(E(\lambda_k) - E(\lambda_{k-1}))(F(\lambda_k^1) - F(\lambda_k^2))\chi\}\|^2 \\
&= \sum_k \|(E(\lambda_k) - E(\lambda_{k-1}))(F(\lambda_k^1) - F(\lambda_k^2))\chi\|^2 \\
&= \|\chi\|^{-2} \sum_k \{\|(E(\lambda_k) - E(\lambda_{k-1}))\chi\|^2 \|(F(\lambda_k^1) - F(\lambda_k^2))\chi\|^2\} \\
&\leq \sup_k \|(F(\lambda_k^1) - F(\lambda_k^2))\chi\|^2 .
\end{aligned}$$

By Lemma2.2 this can be made arbitrarily small by choosing Λ_1, Λ_2 sufficiently fine. Hence $\chi_b = \lim_{\Lambda \subset [0,b]} K(\Lambda)\chi$ exists. By approximating $\chi_{b'}$ (for $b' > b$) using partitions of which b is an element, it is seen by similar arguments that $\|\chi_{b'} - \chi_b\|^2 = \|(E(b') - E(b))\chi\|^2 \to 0$ as $b, b' \to \infty$. Hence $\lim_b \chi_b = \lim K(\Lambda)\chi$ exists.

Lemma2.5 $K(\Lambda)$ converges strongly to a unitary limit $K(x,t)$.

Proof For Λ a partition of $[0,b]$, $K(\Lambda)^* K(\Lambda) = E(b)$. Hence $K(\Lambda)$ is a contraction. Hence strong convergence on finite linear combinations of elements of S, which follows from Lemma 2.4, implies strong convergence everywhere. Also $\lim K(\Lambda)$ is isometric. Since $K(\Lambda)^*$ is obtained from $K(\Lambda)$ by replacing x by $-x$, $K(\Lambda)^*$ also converges to an isometric limit. From this it follows that both limits are unitary.

Proof of Theorem 2.1 For $\chi_1, \chi_2 \in S$ we have, using Lemma2.3,

$$\begin{aligned}\langle K(x,t)\chi_1,\chi_2\rangle &= \lim \langle K(\Lambda)\chi_1,\chi_2\rangle \\ &= \lim \sum_k\{\langle\chi_1,\chi_2\rangle^{-1}\langle F(\lambda_k)\chi_1,\chi_2\rangle\langle(E(\lambda_k)-E(\lambda_{k-1}))\chi_1,\chi_2\rangle\} \\ &= \int_0^\infty \langle\chi_1,\chi_2\rangle^{-1}\langle e^{ixP(t+\lambda)}e^{-ixP(\lambda)}\chi_1,\chi_2\rangle d\langle E(\lambda)\chi_1,\chi_2\rangle,\end{aligned}$$

showing that for fixed t $\langle K(.,t)\chi_1,\chi_2\rangle$ is measurable. Since S is total we conclude that K(.,t) is weakly measurable. That K(.,t) is a one-parameter group follows from the corresponding property for each $\lambda\geq 0$ of $x\mapsto e^{ixP(t+\lambda)}e^{-ixP(\lambda)}$ by approximating by Riemann-Stieltjes sums. Since a weakly measurable one parameter unitary group is necessarily strongly continuous, the existence of $P_T(t)$ is established. The argument for $Q_T(t)$ is completely analogous. That P_T, Q_T satisfy the defining properties 0),1) of a canonical Wiener process follows again using Riemann-Stieltjes sum approximations from the corresponding properties of the process $P_\lambda(t)=P(\lambda+t)-P(\lambda)$, $Q_\lambda(t)=Q(\lambda+t)-Q(\lambda)$.

3. <u>The strong Markov property</u>

The <u>post-T algebra</u> is the von Neumann algebra ${}_TN$ generated by the spectral projections of $P_T(t), Q_T(t), t\geq 0$. The <u>pre-T algebra</u> is the von Neumann algebra N_T comprising those operators $A\in N$ such that for all $\lambda\geq 0$

$$AE(\lambda) = E(\lambda)A \in N_\lambda. \qquad (3.1)$$

<u>Theorem 3.1</u> The von Neumann algebras ${}_TN, N_T$ are independent and (P_T, Q_T, Ψ) is a canonical Wiener Process.

<u>Proof</u> For $A \in N_T$ we have

$$\begin{aligned}Ae^{ixP_T(t)} &= A \lim \sum_k F(\lambda_k)(E(\lambda_k)-E(\lambda_{k-1})) \\ &= \lim \sum_k A(E(\lambda_k)-E(\lambda_{k-1}))F(\lambda_k) \\ &= \lim \sum_k F(\lambda_k)\ A(E(\lambda_k)-E(\lambda_{k-1})) \\ &= \lim \sum_k F(\lambda_k)(E(\lambda_k)-E(\lambda_{k-1}))A \\ &= e^{ixP_T(t)}A.\end{aligned}$$

Similarly A commutes with each $e^{ixQ_T(t)}$. Since these operators generate ${}_TN$ A commutes with all elements of ${}_TN$. For each subinterval $\Delta = (a,b]$ of $R_{>0}$ we write

$$p_{T\Delta} = P_T(b)-P_T(a), \quad q_{T\Delta} = Q_T(b)-Q_T(a), \quad \Delta+\lambda = (a+\lambda, b+\lambda].$$

Then, using (3.1) and Lemma (2.3), if $\Delta_1,\Delta_2,\dots,\Delta_n$ are disjoint subintervals, we have

$$\begin{aligned}&\langle A\exp(i\sum_j(x_j p_{T\Delta_j}+y_j q_{T\Delta_j}))\Psi,\Psi\rangle \\ &= \lim \sum_k\{\langle A(E(\lambda_k)-E(\lambda_{k-1}))\ \exp(i\sum_j(x_j p_{\Delta_j+\lambda_k}+y_j q_{\Delta_j+\lambda_k}))\Psi,\Psi\rangle\} \\ &= \lim \sum_k\{\langle A(E(\lambda_k)-E(\lambda_{k-1}))\Psi,\Psi\rangle\langle \exp(i\sum_j(x_j p_{\Delta_j+\lambda_k}+y_j q_{\Delta_j+\lambda_k}))\Psi,\Psi\rangle\}.\end{aligned}$$

From the fact that for each $\lambda\geq 0$, $(P_\lambda, Q_\lambda, \Psi)$ is a canonical Wiener process, where $P_\lambda(t)=P(t+\lambda)-P(\lambda)$, $Q_\lambda(t)=Q(t+\lambda)-Q(\lambda)$, it follows that for each k

$$\langle \exp(i\sum_j(x_j p_{\Delta_j+\lambda_k}+y_j q_{\Delta_j+\lambda_k}))\Psi,\Psi\rangle = \langle \exp(i\sum_j(x_j p_{\Delta_j}+y_j q_{\Delta_j}))\Psi,\Psi\rangle.$$

Using the fact that

$$\lim \sum_k \langle A(E(\lambda_k)-E(\lambda_{k-1}))\Psi,\Psi\rangle = \langle A\Psi,\Psi\rangle$$

we obtain

$$\langle A \exp(i\sum_j(x_j p_{T\Delta_j}+y_j q_{T\Delta_j}))\Psi,\Psi\rangle = \langle A\Psi,\Psi\rangle\langle \exp(i\sum_j(x_j p_{\Delta_j}+y_j q_{\Delta_j}))\Psi,\Psi\rangle. \quad (3.2)$$

Setting A to be the identity in (3.2) shows that (P_T, Q_T, Ψ) is a canonical Wiener process. Since linear combinations of operators of the form $\exp(i\sum_j(x_j p_{T\Delta_j}+y_j q_{T\Delta_j}))$ are weakly dense in ${}_T N$, (3.2) also shows that ${}_T N, N_T$ are independent.

References

[1] Cockroft, A.M., Gudder, S.P., Hudson, R.L.: A quantum-mechanical functional central limit theorem. J. multivariate Anal. 7, 125-148 (1977).

[2] Cockroft, A.M., Hudson, R.L.: Quantum-mechanical Wiener processes. J. multivariate Anal. 7, 107-124 (1977).

[3] Cushen, C.D., Hudson, R.L.: A quantum-mechanical central limit theorem. J. appl. Prob. 8, 454-469 (1971).

[4] Guichardet, A.: Symmetric Hilbert spaces and related topics. Lecture Notes in Mathematics, Vol. 261. Berlin, Springer (1972).

[5] Hunt, G.: Some theorems concerning Brownian motion. Trans. Amer. Math. Soc. 81, 294-319 (1956).

RANDOM LINEAR FUNCTIONALS AND WHY WE STUDY THEM

Marek Kanter
Sir George Williams Campus
Concordia University
Montreal, Canada

1. Introduction

In this note we present a particular result about the representation of random linear functionals on a Levy process by means of a stochastic integral. For motivation we precede this result by comments of a more general nature.

The subject we treat is the theory of measures on a linear space S. Our point of view is axiomatic in the sense that we study the interplay between measure theoretic notions and the linear structure of the space S. However our point of view is also fruitful for applications because all real valued stochastic process induce a measure on some linear space S (this fundamental result is due to Kolmogorov). It turns out that linear space considerations can yield interesting results for particular stochastic processes in a simple and efficient way (see e.g. [2], [3], or [10]).

Let us formulate in more detail the general problem we are treating. We suppose (S,A) is a measurable linear space, i.e. that A is a σ-field of subsets of S such that addition is AxA measurable and scalar multiplication is $B_R xA$ measurable where B_R stands for the Borel subsets of R, the reals. We suppose also that we are given a collection $M_0 = (\phi_t; t\varepsilon T)$ of real linear functionals defined on S such that A is equal to the σ-field generated by M_0. For example if $S=R^\infty$ on $R^{[0,b)}$ then we let M_0 be the collection of coordinate evaluations. Suppose now that μ is a probability measure on (S,A); then the functions $(\phi_t; t\varepsilon T)$ are random variables and jointly define a stochastic process. We will call (S,A) a linear sample space for the stochastic process $(\phi_t; t\varepsilon T)$ and we will say that $(\phi_t; t\varepsilon T)$ induces the measure μ on (S,A). We will denote by A_μ the μ completion of A.

Defn. 1.1 We define M_1, the random linear functionals on (S,A,μ) to be the set of all A_μ measurable, real valued, linear functionals on S.

We now turn to the question : why study M_1 ? In fact, the most general problem of interest to probabilists is the analysis of the structure of M_∞, the set of all A_μ measurable real valued functions on S. (Note that if the joint distributions of the random variables $(\phi_t; t\epsilon T)$ are given then the structure of M_∞ will not depend on the particular probability space where they live. On the other hand the structure of M_1 will depend on the particular linear sample space S upon which the linear functionals ϕ_t are defined). Since the analysis of M_∞ is an impossible goal in general we propose the study of M_1 (for any linear sample space of the stochastic process $(\phi_t; t\epsilon T)$) as a non-trivial first step. In fact if we let $\overline{M}_0$ be the closure in the sense of convergence in probability of finite linear combinations of elements in M_0 then we have,

$$(1.1) \qquad \overline{M}_0 \subset M_1 \subset M_\infty ,$$

where we note that $\overline{M}_1 = M_1$.

It turns out that the inclusion $\overline{M}_0 \subset M_1$ may be strict (as shown in [8]). In [15] Shorokhod claims and falsely proves that $\overline{M}_0 = M_1$ in the case $S=H$, a Hilbert space, and M_0 is the class of continuous linear functionals on H (however the counter example in [8] is easily transferable to Hilbert spaces).

We conclude that the study of $\overline{M}_0$ is in fact preliminary to the study of M_1. The following example shows that $\overline{M}_0$ is itself not entirely accessible.

Example 1.1 Let $S=R$, $A=B_R$, $M_0 =$ all finite linear combinations $\sum_{-M}^{M} c_k e^{ikx}$ with $k\neq 0$, and $\mu =$ Lebesgue measure on $[-\pi, \pi]$. Then $\overline{M}_0$ contains all Borel measurable real valued functions defined on R, with period 2π (see Goffman [4]). However it is an interesting challenge to actually construct a sequence of elements in M_0 which converge in measure to 1.

We now define a class of measures which we find of interest.

Defn. 1.1 Let μ be a probability measure on the measurable linear space (S,A). We say that μ is linearly injective if for any measurable linear space (S',A') and any linear transformation $U : S \to S'$ with $B = U^{-1}(A') \subset A_\mu$ the condition $B_\mu = A_\mu$ implies that there exists a linear transformation $V : S' \to S$ with $V^{-1}(A) \subset A'_{\mu'}$ and $\mu = \mu' \circ V^{-1}$, where $\mu' = \mu \circ U^{-1}$.

We shall show in the appendix that not all probability measures are injective, even when $S=R^2$.

In the following example we show an interesting consequence of the property of linear injectivity for a measure.

<u>Example 1.3</u> Let $(X_n;n\geq 0)$ be a sequence of random variables which induce the measure μ on R^∞ and suppose μ is linearly injective. Let $U : R^\infty \to R^\infty$ be defined by setting $U(X_0,X_1, \ldots, X_n, \ldots) = (X_1,X_2, \ldots, X_n, \ldots)$. The condition that $B_\mu = A_\mu$ here is interpreted as

<u>Condition (A)</u> X_0 is measurable with respect to the σ-field generated by $(X_1, \ldots, X_n, \ldots)$ when completed by throwing in the null sets generated by $(X_0,X_1, \ldots, X_n, \ldots)$.

The existence of V in Defn. 1.1 here simply means;

<u>Conclusion (B)</u>: There exists a random linear functional f on $(R^\infty, B_{R^\infty}, \mu')$ such that $f=X_0$ a.s. (Here μ' is the measure induced on R^∞ by the process $(X_1,X_2, \ldots, X_n, \ldots)$).

If in (B) we can choose f in $\overline{M}_0$ then we shall call it Conclusion (B^*).

Let us remark that any product measure on R^n is linearly injective. Also for any sample space (S,A) all Gaussian measures μ are linearly injective. Furthermore (A) implies (B^*) for Gaussian measures. It is tempting to conjecture that any product measure is also linearly injective. It is also tempting to conjecture that any symmetric stable measure is linearly injective (see [2] for the definition of such measures), and that (A) implies (B) when Y_k are symmetric stable.

We shall now show that if Y_j are identically distributed, symmetric stable of index $\alpha \in (0,2)$ then (A) does not imply (B^*). This contradicts the main theorem in [9]. However the proof in [9] was based on a theorem of Shorokhod [16]. In [17], Shorokhod has retracted that theorem.

In the following we let $\alpha \in (1,2)$ and set $\beta = \alpha/(\alpha-1)$.

<u>Example 1.4</u> Let $(c_n;n\geq 0)$ be a sequence of numbers with $\sum_0^\infty c_n^2 \leq \infty$ and $\sum_0^\infty |c_n|^\beta$ finite. Assume also that $c_0=0$ and set $X_n = Y_n+c_nY_0$ for $n\geq 0$. The main theorem of [6] (see also [7]) shows that there exists a Borel measurable real

valued function h on R^∞ such that $h(Y_1+sc_1, Y_2+sc_2, \ldots, Y_n+sc_n, \ldots) = s$ a.s. When we randomize the value s according to the distribution of Y_0 we get that $h(X_1, X_2, \ldots, X_n, \ldots) = X_0$ a.s., i.e. that Condition A holds. To see that Conclusion B* does not hold we note that for any sequence $b_1, \ldots, b_m$ of real numbers the random variable $X_0 - \sum_1^m b_i X_i$ is small in probability if and only if the two random variables $(1 - \sum_1^m b_i c_i)Y_0$ and $\sum_1^m b_i Y_i$ are both small in probability. Using characteristic functions we note that $\sum_1^m b_i Y_i$ is small in probability if and only if $\sum_1^m |b_i|^\alpha$ is small. Finally the condition that $\sum_1^\infty |c_i|^\beta < \infty$ implies that $\sum_1^m b_i c_i$ is also small hence, in fact, $(1 - \sum_1^m b_i c_i)Y_0$ cannot be small in probability.

Remark To construct a counter example in the case $0<\alpha\leq 1$ we need only to take a sequence $(c_n;n\geq 0)$ with $\sup_{n\geq 0} |c_n| < \infty$ and $\sum_0^\infty c_n^2 = \infty$; then proceed as above.

For symmetric stable processes of index $\alpha\varepsilon(0,2]$ the study of $\overline{M}_0$ is exactly equivalent to the study of arbitrary closed linear subspace of $L^\alpha[0,1]$. It is an interesting problem to determine which closed linear subspaces of $L^\alpha[0,1]$ correspond to symmetric stable measures μ for which $\overline{L}_0=L_1$. (We have just seen not all subspaces of $L^\alpha[0,1]$ have this property). It has been shown in [8] (see [18] for the case $0<\alpha\leq 1$) that if $\overline{M}_0 = L^\alpha[0,1]$ then $\overline{M}_0=M_1$.

We see from Example 1.4 that (A) does not necessarily imply (B^*) even for symmetrically distributed stochastic processes $(X_n;n\geq 0)$. However in the case when X_n are a stationary moving average process with some extra conditions then (A) does imply (B^*) as shown in [11] (where this qualitative statement is made more precise).

In the next section of this paper we shall study M_1 in the case μ is a measure induced on S by a process $(X(t) ; t\varepsilon[0,b))$ with independent identically distributed increments (such a stochastic process is called a Levy process). S may be taken to be $R^{[0,b)}$ or the set of bounded real valued functions on $[0,b)$ with no discontinuities of the second kind. In this paper it will be convenient to let $S = L^2[0,b)$ and to let M_0 be the set of continuous linear functionals on $L^2[0,b)$. (In the above b may be finite or $+\infty$).

In [19] it is posed as a conjecture if every random linear functional on (S,A,μ) is in fact a random integral, where $S = L^2[0,b)$, A is the σ-field generated by M_0 , and μ is the measure induced on S by a Levy process with no Gaussian part. However we had already answered this problem affirmatively in a previous paper [5], unpublished (now to appear in Colloq. Math). This shows, in effect, that $\bar{M}_0=M_1$ for this case. In Section 3 of this paper we present an abridged version of the results in [5]. Furthermore, to save space, we shall make use of results proved by Urbanik in [19] wherever possible, rather than to go through similar arguments in [5]. We shall assume that all Levy processes we treat as symmetrically distributed for simplicity.

The problem we treat in Section 3 is linked in an interesting way to the conjecture that the measure μ induced by a Levy process on S is linearly injective. Let $(X(t)\ ;\ t\varepsilon [0,b))$ be a Levy process. Let $X(t) = Y(t)+Z(t)$ be its decomposition into Gaussian and non-Gaussian part respectively (see e.g. Loeve [13]). Consider the map U from SxS into S sending the vector process $(Y(t), Z(t))$ into $(Y(t) + Z(t))$. In fact we have $B_\mu=A_\mu$ for this map, using our previous notation and letting μ stand for the measure induced on SxS by the process $(Y(t), Z(t))$. (This in effect just amounts to saying that $Y(t)$ and $Z(t)$ are measurable with respect to the completed σ-field generated by $X(t)$) . If μ were linearly injective then we would conclude that there exists a linear map $V : S\to SxS$ such that $V(X(t)) = (Y(t), Z(t))$ a.s., i.e. that the Levy decomposition can be achieved in a linear way. In fact Levy in his paper [12] seems to assume this is the case, however we have been unable to rigorously verify this except in the easy situation when the non-Gaussian part $Z(t)$ has its jumps absolutely summable over every finite interval (in that case $Z(t)$ is defined as a linear function of $X(s)$ by just adding up the jumps of $X(s)$ in $[0,t)$) .

Borell in [1] also deals with similar problems. In this paper he puts geometric hypotheses on the measure μ , rather than structural probabilistic hypotheses of stability or infinite divisibility, as we do.

Vershik in [21] has a similar point of view to ours. In fact if (S,A,μ) is any measurable linear space which is also a Lebsegue space then (without assuming the existence of a set M_0 as we do) he claims to prove that the collection M_1 of random linear functionals on S has the property that the σ-field generated by M_1 is equal to A_μ . However his proof seems incomplete to us and we consider this problem still open. (For instance if $S = L^\alpha[0,1]$ with $\alpha\varepsilon(0,1)$ and if μ is an arbitrary measure on (S,A) , where A consists of the Borel subsets of S , then we cannot even show that there exists any f in M_1 other than $f\equiv 0$) .

2. Random Integrals and Levy Measures

Let $(X(t)\ ;\ t \in [0,b))$ be a symmetric Levy measure (where $b>0$ may be finite or infinite). Assume that $X(t)$ has no Gaussian component. We let $\phi_t(u) = E(e^{iuX(t)})$, the characteristic function of $X(t)$. We then have

$$(*) \qquad \log \phi_t(u) = t \int_0^\infty (\cos (uy) - 1)\, dM(y)$$

where M is a non-negative measure on $(0,\infty)$ with $\int_0^\infty (y^2/1+y^2)\, dM(y) < \infty$. If $\tilde{h}$ is a Borel measurable real valued function on $[a,b]$ it is shown in [12, p.243] that $\int_a^b \tilde{h}(t)\, dX(t)$ exist a.s. and is a linear functional of the jumps of the $X(t)$ process if and only if $\int_0^b \int_0^1 \min (1, (\tilde{h}(t)\, y)^2\, dydt < \infty$. By an easy computation this last condition is seen to be equivalent to the condition that $\int_0^b \Psi(|\tilde{h}(t)|)dt < \infty$ where Ψ is as given in [19, p. 258]. If either of these conditions hold we shall write $\tilde{h} \in L(\mu)$, where μ is the measure induced on (S,A) by the process X. (Note we are taking $S = L^2\,[a,b]$ henceforth).

In summary we see that if $\tilde{h} \in L(\mu)$ then the functional $x \to \int_0^b \tilde{h}(t)\, dx(t)$ exists on a linear subspace of S with μ measure 1, hence $h(x) = \int_a^b \tilde{h}(t)\, dx(t)$ defines a random linear functional on (S,A,μ). In the next section we shall prove the converse.

3. Main Result

The following simple lemma is crucial.

Lemma 3.1 Let f be a random linear functional on (S,A,μ) where (S,A) is any measurable linear space and where $\mu = \mu_1 * \mu_2$, the convolution of two probability measures on (S,A). Then f is random linear functional on (S,A,μ_i) for $i = 1,2$. Furthermore $\phi = \phi_1\phi_2$ where ϕ_i is the characteristic function of the random variable f defined on (S,A,μ_i), while ϕ is the characteristic function of f on (S,A,μ).

Proof For r rational let $D_r = \{x;\ x \in S\ ,\ f(x) < r\}$. Since f is A_μ measurable there exist $A_r, B_r \varepsilon A$ such that $A_r \subset D_r \subset B_r$ and $\mu(B_r/A_r) = 0$. Now

$$\mu(B_r/A_r) = \int_S \mu_1((B_r/A_r) - y)\, d\mu_2(y)$$

by Fubini's Theorem. We let $C_r = \{y;\ \mu_1((B_r/A_r) - y) = 0\}$ and conclude $C_r \varepsilon A$ with $\mu_2(C_r) = 1$. If we let $C_r = \bigcap_{\text{rational}} C_r$ it still follows that $\mu_2(C_r) = 1$, hence there exists $x_0 \varepsilon C$. It then follows that $\mu_1(B_r/A_r) - x_0) = 0$ for all r rational. This immediately implies that the function f_{x_0} defined by $f_{x_0}(x) = f(x+x_0)$ is A_{μ_1} measurable. Now $f = f_{x_0} - f(x_0)$ hence f is also A_{μ_1} measurable. Reversing the roles of μ_1 and μ_2 we conclude that f is μ_2 measurable.

The last statement of the lemma has straightforward verification.

q.e.d.

Before preceding to our main theorem we need some further notation. If $X(t)$ is a Levy process on $[0,b)$ and M is the Levy measure for X as defined in (*) , we define for $\gamma>0$ the measures $M_\gamma(A) = M(A \cap (\gamma,\infty))$. We let X_γ and X^γ denote the Levy processes associated with M_γ and M^γ respectively, and we let μ_γ, μ^γ denote the measures induced on (S,A) by X_γ and X^γ respectively. It is clear that $\mu = \mu_\gamma * \mu^\gamma$.

Theorem 3.1 Let X be a symmetric Levy process on $[0,b)$ with no Gaussian part. Let f be a random linear functional on (S,A,μ) where $S = L^2[0,b)$. Then there exists a Borel measurable function $\tilde{f}$ in $L(\mu)$ such that

$$f(x) = \int_0^b \tilde{f}(t)\, dx(t) \quad (\mu \text{ a.s.})$$

Proof Assume first that $b<\infty$. Let $\gamma>0$ and write $\mu = \mu_\gamma * \mu^\gamma$ as above. By Lemma 3.1 we know that f is a random linear functional on (S,A,μ_γ) hence by Theorem 3 of [19] we know there exists a Borel measurable function $\tilde{f} \varepsilon L(\mu)$ with $f(x) = \int_0^b \tilde{f}(t)\, dx(t)$ (μ_γ a.s.). We may choose $\gamma_0>0$ so small that μ_{γ_0} is not concentrated at $0 \varepsilon S$, (otherwise our process X is identically zero). Then the argument in Theorem 3 of [19] show that $\tilde{f}$ is uniquely specified up to sets of Lebesgue measure 0 and that $\tilde{f}$ does not vary with γ for $\gamma \varepsilon (0,\gamma_0]$.

We now show that $\tilde{f} \varepsilon L(\mu)$. Let $\gamma_{-1} = +\infty$, $\gamma_n = n^{-1}\gamma_0$ for $n \geq 1$, and let μ_n be the measure induced on (S,A) by the Levy process with Levy measure

$M_n(A) = M(A \cap (\gamma_n, \gamma_{n-1}])$ for $n=0,1,2,\ldots$. Let W_n be a sequence of independent random variables each of which is distributed like $\int_0^b \tilde{f}(t)\,dx(t)$ on the probability space (S,A,μ_n) . Let ϕ stand for the characteristic function of the random variable f on (S,A,μ) . Then, using Lemma 3.1, we see that for any n we have $\phi = \phi_n \Psi_n$ where ϕ_n is the characteristic function of $\sum_0^n W_i$ while Ψ_n is the characteristic function of f on (S,A,μ^{γ_n}) . We conclude from [13, Lemma 37.Vb] that $\sum_0^n W_i$ converges a.s. A standard argument (see e.g. [14]) now show that $\tilde{f} \in L(\mu)$.

Since $\tilde{f} \in L(\mu)$ we know there exists a random linear function f^* on (S,A,μ) such that $f^*(x) = \int \tilde{f}(t)\,dx(t)$ (μ a.s.). We now show that $g \equiv f - f^*$ vanishes μ a.s. Consider the product probability space $(\hat{S},\hat{A},\hat{\mu}) = (\bigotimes_{k=0}^{\infty} S_k, \bigotimes_{k=0}^{\infty} A_k, \bigotimes_{k=0}^{\infty} \mu_k)$ where $S_k = S$ and $A_k = A$ for all k . For $\hat{x} = (x_0, x_1, \ldots, x_n, \ldots) \in \hat{S}$ let $\pi_n(\hat{x}) = \sum_0^n x_i$. We claim that $\lim_{n\to\infty} \pi_n(\hat{x})$ converges $\hat{\mu}$ a.s. (as a sequence in $S = L^2[a,b))$. In fact for any $y \in S$ we get that $(y.\pi_n(\hat{x}))$ converges a.s. by arguing as above using [13, Lemma 37.Vb] which is enough to prove the claim since $||\pi_n(\hat{x})||^2 \geq ||\pi_{n-1}(\hat{x})||^2$. (Here $(y.z) = \int_0^b y(t)\,z(t)\,dt$ for $y,z \in S$ and $||y||^2 = (y.y))$.

Let $H = \{\hat{x};\ \hat{x} \in S'$ and $\lim_{n\to\infty} \pi_n(\hat{x})$ exists $\}$. We define $\pi(\hat{x}) = \lim_{n\to\infty} \pi_n(\hat{x})$ for $\hat{x} \in H$ and $\pi(\hat{x}) = 0$ for $\hat{x} \notin H$. We let $\pi^n(\hat{x}) = \pi(\hat{x}) - \pi_n(\hat{x})$ for $\hat{x} \in \hat{S}$. We have $g(\pi_n(\hat{x})) = 0$ ($\hat{\mu}$ a.s.) by choice of f . It follows that $g(\pi(\hat{x})) = g(\pi^n(\hat{x}))$ ($\hat{u}$ a.s.) for all n . This shows that $g \circ \pi$ is measurable with respect to $\bigcap_{m=1}^{\infty} \bigotimes_{k=m}^{\infty} A_k$, hence $g \circ \pi$ is constant ($\hat{\mu}$ a.s.) by the Kolmogorov 0-1 law. Now $g \circ \pi$ on $(\hat{S},\hat{A},\hat{\mu})$ has the same distribution as g on (S,A,μ) which shows that $g=0$ (μ a.s.) since, of course, g has a symmetric distribution.

We conclude that $f(x) = \int_0^b \tilde{f}(t)\,dx(t)$ (μ a.s.) if b is finite. If $b=\infty$ define for each $s>0$ the stochastic process $(Y_s(t);\ t\geq 0)$ by $Y_s(t) = X(s \wedge t)$ and define the stochastic process $(Y^s(t);\ t\geq 0)$ by $Y^s(t) = X(t) - X(s\wedge t)$, where $s \wedge t$ stands for the minimum of s,t . If we let ν_s and ν^s be the measures induced on (S,A) by the processes Y_s and Y^s respectively then clearly $\mu = \nu_s * \nu^s$.

Using Lemma 3.1 we get that f is a random linear functional on (S,A,ν_s) . The preceding argument now shows that $f(x) = \int_0^\infty \tilde{f}(t)\, dx(t)$ (ν_s a.s.) for all $s>0$. We now argue as before using a product measure built from the measures ν_s and the Kolmogorov 0-1 law to conclude that $f \in L(\mu)$ and that $f(x) = \int_0^\infty \tilde{f}(t)\, dx(t)$ (μ a.s.) .

q.e.d.

References

1. C. Borell, Random linear functionals and subspaces of probability one, Arkiv for Matematik 14 (1976), 79-92.
2. R.M. Dudley and M. Kanter, Zero-one laws for stable measures, Proc. Amer. Math. Soc. 45 (1974), 35-47.
3. X. Fernique, Integrability des vecteurs Gaussiens. C.R. Acad. Sci. Paris Ser. A 270 (1970), 1698-1699.
4. C. Goffman and D. Waterman, Some Aspects of Fourier Series, Am. Math. Monthly 77 (1970), 119-133.
5. M. Kanter, Completion measurable linear functionals on a probability space, to appear in Colloq. Math.
6. M. Kanter, On distinguishing translates of measures, Ann. Math. Stat. 49 (1969), 1773-1777.
7. M. Kanter, Correction to "On Distinguishing Translates of Measures", Ann. of Probability 3 (1975), 189-190.
8. M. Kanter, Linear sample spaces and stable processes, Journal of Functional Analysis 9 (1972), 472-474.
9. M. Kanter, On the spectral representation for symmetric stable random variables Z. Wahrscheinlichkeitsth. Verw. Geb. 23 (1972), 1-6.
10. M. Kanter, On the boundedness of stable processes, Transactions of the Seventh Prague Conference on Information Theory (1974).
11. M. Kanter, Lower bounds for non-linear prediction error in moving average processes, submitted.
12. P. Levy, Functions aleatoires à correlation lineaire, Illinois Journal of Mathematics 1 (1957), 217-258.
13. M. Loeve, Probability Theory 3d ed. Van Nostrand, Princeton, N.J. (1963).
14. D. Shale and W.F. Stinespring, Wiener processes II, Journal of Functional Analysis 5 (1970), 375-400.
15. A.V. Skorokhod, Integration in Hilbert Space, Springer-Verlag, Berlin (1974).
16. A.V. Skorokhod, On the density of probability measures in functional spaces, Proc. 5th Berk. Symp. Math. Stat. Prob. Vol. 2 (1967), 163-182.
17. A.V. Skorokhod, On admissible translates of measures in Hilbert spaces, Theory of Probability and Applications 15 (1970), 557-580.
18. M.A. Tortrat, Pseudo martingales et lois stables, C.R. Acad. Sc. Paris. Ser.A . 281 (1975), 463-465.
19. K. Urbanik, Random linear functionals and random integrals, Colloq. Math. 23 (1975), 255-262.

20. K. Urbanik and W.A. Woycynski, A random integral and Orlicz spaces, Bulletin de l' Academic Polonaise des Sciences Mathematiques, Astronomiques, et Physiques 15, (1967), 161-169.

21. A.M. Vershik, Axiomatics of the theory of measure in linear spaces, Dokl. Akad. Nauk SSR 178 (2) (1968), 278-281.

APPENDIX

The following example shows that even measures on R^2 are not necessarily linearly injective.

Example Let $S=R^2$ and let $A=B_{R^2}$, the Borel subsets of R^2 . Let h be any real valued Borel measurable function on R and let μ be any probability measure on R^2 which is concentrated on the set $C = \{(y,x);\ y=h(x)\}$. Let $U : R^2 \to R$ be defined by setting $U(x,y) = x$. The fact that $B_\mu = A_\mu$ is easy to check. Furthermore for any linear transformation $V : R \to R^2$ the measure $\mu' o V^{-1}$ is concentrated on a line passing through the origin, hence $\mu \neq \mu' o V^{-1}$ in general.

CONTROL MEASURE PROBLEM IN SOME CLASSES OF F-SPACES

by

Przemysław Kranz

Institute of Mathematics, Polish Academy of Sciences, Poznań

In this note we present some results related to a problem of Mrs. Maharam [9], i.e. whether the existence of a continuous submeasure on a Boolean algebra implies the existence of a continuous strictly positive measure on it. We characterize this property in F-spaces and in particular in p-convex spaces and we show that L^p-valued measures $(0 < p \leq 1)$ have control measures.

Let $\mathcal{B}$ be a Boolean algebra. A submeasure on $\mathcal{B}$ is a function $\nu : \mathcal{B} \longrightarrow R_+$ with the properties

(i) $\nu(0) = 0, \quad \nu(1) > 0$

(ii) $a \leq b$ implies $\nu(a) \leq \nu(b), \quad a,b \in \mathcal{B}$

(iii) $\nu(a \vee b) \leq \nu(a) + \nu(b)$ for every $a,b \in \mathcal{B}$.

If, instead of (iii) we require that $\nu(a \vee b) = \nu(a) + \nu(b)$ for all $a,b \in \mathcal{B}$, $a \wedge b = 0$ then ν is obviously a (finitely additive) measure. A continuous submeasure is a submeasure ν on a σ-algebra satisfying: (i') $\nu(a) = 0$ if and only if $a = 0$, (ii), (iii), and (iv) if $\{a_n\}$ is a decreasing sequence in $\mathcal{B}$ with $\inf a_n = 0$ then $\nu(a_n) \longrightarrow 0$. A measure $\lambda : \mathcal{B} \longrightarrow R_+$ is called an additive minorant of a submeasure ν if $\lambda \leq \nu$.

If a submeasure ν has a nontrivial additive minorant we say that ν is non-pathological [3], otherwise call ν pathological.

Let $a \in \mathcal{B}$. We say ([7], Lemma 2.2.) that the elements $a_i \in \mathcal{B}$ $(i = 1,\dots,n)$ cover a m-times (denoted $ma = \sum_{i=1}^{n} a_i$) if there exists a double indexed sequence $\{a_{ij}\}_{i=1,j=1}^{n\quad m}$ of elements of such that $a_i = \bigvee_{j=1}^{n} a_{ij}$ $(i = 1,\dots,n)$ and $a = \bigvee_{i=1}^{n} a_{ij}$ for every $j = 1,\dots,m$; where for every i a_{ij}'s are pairwise disjoint in j and for every j a_{ij}'s are pariwise disjoint in i.

If a_{ij}'s of the above decomposition are such that $\bigvee_{i=1}^{n} a_{ij} \leq a$ for every $j = 1,\dots,m$ we say that $\{a_i\}$ cover a at least m times $(ma \leq \sum_{i=1}^{n} a_i)$. A submeasure ν on a Boolean algebras is multiply

<u>subadditive</u> if $m\nu(a) \leq \sum_{i=1}^{n} (a_i)$ whenever $ma \leq \sum_{i=1}^{n} a_i$.

A submeasure ν on a Boolean algebra is multiply subadditive if and only if it is a supremum of some family of measure on $\mathcal{B}$ ([7], Theorem 4). In other words, every nontrivial multiply submeasure is nonpa thological.

Definition.

A submeasure $\nu : \mathcal{B} \longrightarrow R_+$ is an <u>M-submeasure</u> if, for every $a \in \mathcal{B}$, $\nu(a) > 0$, its restriction to the principal ideal $a \wedge \mathcal{B} = \{a \wedge b : b \in \mathcal{B}\}$ is non-pathological. The following is proved in [10], Theorem 2 (see also [6]).

Theorem 1.

A submeasure ν on $\mathcal{B}$ is an M-submeasure if and only if for every $a \in \mathcal{B}$ such that $\nu(a) > 0$ there is a constant $\alpha_o > 0$ such that

$$\alpha_a m\nu(a) \leq \sum_{i=1}^{n} \nu(a_i) \quad \text{whenever} \quad ma \leq \sum_{i=1}^{n} a_i .$$

Theorem 2.

Let ν be a submeasure on a Boolean algebra $\mathcal{B}$. The function

$$\nu^*(a) = \inf \{ \sum_{i=1}^{n} m^{-1} \nu(a_i) : ma \leq \sum_{i=1}^{n} a_i \}$$

is a multiply subadditive minorant of ν.

Proof.

Straightforward verification (for details see [6], Theorem 4).

Theorem 3.

For a submeasure ν defined on a Boolean algebra $\mathcal{B}$, $\nu^*(a) > 0$ if and only if there exists a constant $\alpha_a > 0$ such that

$$(*) \qquad \alpha_a m\nu(a) \leq \sum_{i=1}^{n} \nu(a_i) \quad \text{whenever} \quad ma \leq \sum_{i=1}^{n} a_i ,$$

that is, if and only if is an M-submeasure.

Proof.

Obviously, if such $\alpha_a > 0$ exists then (*) is satisfied, and so $0 < \alpha_a < \inf\{ \sum_{i=1}^{n} m^{-1} \nu(a_i) \} = \nu^*(a)$.

On the other hand assume that no $\alpha_a > 0$ satisfying (*) exists. Then for every $\alpha > 0$ there is $\{a_i\}_{i=1}^{n_\alpha}$ and m_α, n_α such that $\alpha_a m_\alpha \nu(a) > \sum_{i=1}^{n} \nu(a_i)$. Then clearly $\nu^*(a) = 0$, a con tradiction.

Corollary.

Assume that ν is a strictly positive submeasure (i.e.

$\nu(a) = 0 \Longrightarrow a = 0$) on a Boolean algebra $\mathcal{B}$. ν^* is a strictly positive subadditive minorant of ν if and only if, ν is an M-submeasure.

If is multiply subadditive then $\nu = \nu^*$. In fact, $\nu^* \leq \nu$ and conversely, if ν is multiply subadditive then $\nu(a) \leq \sum_{i=1}^{n} m^{-1}\nu(a_i)$ for $ma \leq \sum_{i=1}^{n} a_i$, and so $\nu \leq \nu^*$. More generally, for any M-submeasure ν, if $\nu(a) > 0$ then

(**) $\quad 0 < \nu(a)\alpha_a \leq \nu^*(a) \leq \nu(a)$, where α_a is the same as in (*).

A submeasure ν_1 is called absolutely continuous with respect to a submeasure ν_2 on a Boolean algebra $\mathcal{B}$ ($\nu_1 \ll \nu_2$) if, for every $\varepsilon > 0$ there is a $\delta > 0$ such that $\nu_2(a) < \delta$ implies $\nu_1(a) < \varepsilon$, $a \in$.

ν_1 is equivalent to ν_2 ($\nu_1 \sim \nu_2$) if $\nu_1 \ll \nu_2$ and $\nu_2 \ll \nu_1$.

A submeasure ν is said to be exhaustive if, for every pairwise disjoint sequence $\{a_n\}_{n=1}^{\infty}$ in $\mathcal{B}$, $\nu(a_n) \longrightarrow 0$

Theorem.

Let $\mathcal{B}$ be a Boolean algebra and ν an exahustive and countably subadditive strictly positive submeasure on $\mathcal{B}$. Then $\mathcal{B}$ admits a strictly positive exhaustive additive measure if and only if ν is an M-submeasure:

Proof.

From (**) and 2 it follows that ν is equivalent to ν^* if and only if ν is an M-submeasure. Clearly ν^* is exhaustive and countably subadditive if and only ν is. But if ν is an exhaustive M-submeasure then by Theorem 4 of [10] there exists a strictly positive minorant λ of ν^* which is also an additive minorant of ν. If ν is continuous then so is λ, and exhaustivity of ν implies the exhaustivity of λ.

For the converse assume that ν is not an M-submeasure. Then for some $a \in \mathcal{B}$ its restriction to the Boolean algebra $a \wedge \mathcal{B}$ is pathological. By a Theorem 2 of [3] there exists no measure λ on such that $\lambda \leq \nu$, and so there is no strictly positive exhaustive and countably additive measure on $\mathcal{B}$.

Corollary.

Let $\mathcal{B}$ be an atomless σ-algebra with a continuous submeasure ν. Then $\mathcal{B}$ admits a strictly positive σ-additive measure λ if and only if ν is an M-submeasure.

Proof.
If ν is an M-submeasure which is continuous then there is a strictly positive measure $\lambda \leq \nu$ on $\mathfrak{B}$ which is countably additive. Conversely, let λ be a countably additive strictly positive measure on $\mathfrak{B}$. Then by a preceding theorem ν has to be an M-submeasure.

7. Corollary.
Let ν be as in the previous theorem. Then ν is not an M-submeasure (does not have the Maharam.Property) if and only if there is some $a \in \mathfrak{B}$ with

$$\inf \{ \sum_{i=1}^{n} \frac{\nu(a_i)}{m} : ma \leq \sum_{i=1}^{n} a_i \} = 0$$

The above formula allows construction a pathological submeasure ν, which is exhaustive on a large subcollection $\mathcal{C}$ of elements of $\mathfrak{B}$. It is not, however, known if it is possible to show that an extension of ν to a Boolean algebra generated by $\mathcal{C}$ remains exhaustive, although this extension is, of course, pathological.
The first example of a pathological submeasure was given by W.Herer and J.P.R.Christensen in [3]. Popov also gives a simple example of a pathological submeasure [10].
It is not clear however, if the submeasure in this example is exhaustive.
Suppose that E is an F-space (complete metric linear space). If $\mu : \mathfrak{B} \longrightarrow E$ is a vector measure then the semivariation $\|\mu\|$ defined as

$$\|\mu\|(P) = \sup\{ \|\mu(Q)\| : Q \leq P, \ Q \in \mathfrak{B}\}$$

is a continuous submeasure.
A finite positive measure λ is a control measure for μ if λ is equivalent to $\|\mu\|$
The following question is equivalent to Maharam's problem [4].
Does every measure $\mu : \mathfrak{B} \longrightarrow E$ where E is an F-space have a control measure?
Applying the result of theorem 2 we get that an E-valued measure has a control measure if and only if:

$$\alpha_a \, m \sup_{Q \leq P} \|\mu(Q)\| \leq \sum_{i=1}^{n} \sup_{Q_i \leq P_i} \|\mu(Q_i)\|$$

whenever $m\,P \leq \sum_{i=1}^{n} P_i$ and $Q_i = Q \cap P_i$ $(i = 1,2,\ldots,n)$.

If E is a p-convex F-space, that is the norm $\|\cdot\|$ of E satisfies the following conditions:

(1) $\|x\| = 0 \Longleftrightarrow x = 0$

(2) $\|x+y\| \leq \|x\| + \|y\|$

(3) $\|tx\| \leq \|x\|$, $|t| \leq 1$

(4) $\|\xi_n x\| \longrightarrow 0$ as $\xi_n \longrightarrow 0$

(5) $\|\xi x\| = |\xi|^p \|x\|$ $0 < p \leq 1$.

then a sufficient condition in order that a measure $\mu : \mathcal{B} \longrightarrow E$ has a control measure is:

$$\|\sum_{i=1}^{n} \mu(Q_i)\| \leq \alpha \frac{m^p}{m} \sum_{i=1}^{n} \|\mu(Q_i)\| \quad \text{whenever} \quad mP \leq \sum_{i=1}^{n} P_i;\ \alpha > 0.$$

Note that for $p = 1$ this condition is trivially satisfied, and this is then the classical theorem of Bartle, Dunford and Schwartz [1]. Assume now that μ is a $L^p(X,\Sigma,\gamma)$ valued measure, where $0 < p \leq 1$. As it is well known ([8], 49p. ex. (vi)) these spaces are Riesz p-convex F-spaces, where $\|x\| = \int_X |x(t)|^p d\gamma$.

Let $\mu : \longrightarrow L^p$. Then μ can be decomposed as $\mu = \mu^+ - \mu^-$, where $\mu^+,\mu^- \geq 0$ with respect to the ordering of L^p.
Since μ^+ is a vector measure with the range in the positive cone of L^p it satisfies the property:

$$\mu^+(Q_1 \cup Q_2) + \mu^+(Q_1 \cap Q_2) = \mu^+(Q_1) + \mu^+(Q_2)$$

for all $Q_1, Q_2 \in \mathcal{B}$.
Clearly $\mu^+(Q_1 \cup Q_2) \geq \mu^+(Q_1)$, $\mu^+(Q_2) \geq \mu^+(Q_1 \cap Q_2)$.
It is an easy matter to show that if a,b,c,d are real nonnegative numbers such that $a \geq b$, $c \geq d$ and $a + d = b + c$ then $a^p + d^p \leq b^p + c^p$ for $0 < p \leq 1$.
Therefore we have:

$$(\mu^+)^p(Q_1 \cup Q_2) + (\mu^+)^p(Q_1 \cap Q_2) \leq (\mu^+)^p(Q_1) + (\mu^+)^p(Q_2)$$

this implies that

$$\|\mu^+(Q_1 \cup Q_2)\| + \|\mu^+(Q_1 \cap Q_2)\| \leq \|\mu^+(Q_1)\| + \|\mu^+(Q_2)\|$$

and, as the norm of L^p is monotone

$$\nu^+(P_1 \cup P_2) + \nu^+(P_1 \cap P_2) \leq \nu^+(P_1) + \nu^+(P_2);$$

where $\nu^+(P) = \sup_{Q \leq P} \|\mu^+(Q)\|$.

This shows that ν^+ is a multiply subadditive submeasure, ([2],[5]) and therefore admits a control measure λ^+.

By a similar argument μ^- admits a control measure λ^-.

But since $0 = \nu(A) \Longleftrightarrow |\mu|(A) = 0 \Longleftrightarrow \lambda^+(A) + \lambda^-(A) = 0$ this shows that $\lambda \equiv \lambda^+ + \lambda^-$ is a control measure for μ.

Therefore we have proved:

8. Theorem.

Every measure $\mu : \mathfrak{B} \longrightarrow L^p(X,\Sigma,\gamma)$ where $0 < p \leq 1$ admits a control measure.

Acknowledgments.

I express my thanks to Dr Alain Costé of Caen University for his thorough help in preparation of this note.

References:

1. R.G.Bartle, N.Dunford and J.T.Schwartz; Weak compactness and vector measures, Canad. J. Math. 7(1955), p. 289-305.
2. B.J.Eisenstatt, G.G.Lorentz; Boolean rings and Banach latties, Illinois J.Math. 3(1959), p. 524-531.
3. W.Herer, J.P.R.Christensen; On the existence of pathological submeasures and the construction of exotic topological groups, Math. Ann. 213(1975), p. 203-210.
4. N.J.Kalton; Linear operators whose domain is locally convex (preprint).
5. P.Kranz; Sandwich and extension theorems on semigroups and lattices, Comm. Math. (Prace Mat.) XVIII (1975), p. 193-200.
6. ——; Submeasures on Boolean algebras and applications to control measure problem, ibidem. XXII (to appear).
7. G.G.Lorentz, Multiply subadditive functions, Canad. J. Math. 4 (1952), p. 455-462.

8. W.A.J.Luxemburg, A.C.Zaanen; Riesz Spaces I, Amsterdam 1971.
9. D.Maharam, An algebraic characterization of measure algebras, Ann. Math. 48(1947), p. 154-167.
10. V.A.Popov, Additive and subadditive functions on Boolean algebras (in Russian), Sibirski Math. Zh. 17(1976), p. 331-339.

APPLICATION DES PROPRIÉTÉS DES FONCTIONS PLURISOUSHARMONIQUES A UN PROBLÈME DE MESURE DANS LES ESPACES VECTORIELS COMPLEXES

par P.LELONG (Paris)

Université PARIS VI
Lab."Analyse complexe et Géométrie"
Associé au C.N.R.S. (L.A.213)
Tour 45-46,5e ét., 4,Pl.Jussieu

1. Introduction. Les mesures (essentiellement positives) liées à l'étude des fonctions holomorphes dans un espace vectoriel complexe possèdent des propriétés de régularité exceptionnelles, et semblent pouvoir être étudiées en dehors d'une théorie générale de la mesure dans les espaces de dimension infinie. Des exemples de telles mesures sont donnés en dimension finie par les "aires" des ensembles analytiques complexes, et d'une manière générale par les mesures de Radon construites à partir du courant positif fermé $t = \frac{i}{\pi}d'd''f = \frac{1}{2\pi}\, dd_c f$ où f est plurisousharmonique. En particulier dans C^n, soit $\sigma = t \wedge \beta_{n-1}$ (où β_p est l'élément de volume des sous-espaces C^p, $1 \leqslant p \leqslant n$). La mesure positive σ possède une propriété de régularité remarquable. Si $\sigma(x,r)$ est la mesure σ portée par une boule $B(x,r)$, de centre x, de rayon r, le quotient

$$\nu(x,r) = (\tau_{2n-2}\, r^{2n-2})^{-1}\, \sigma(x,r)$$

où τ_p est le volume de la boule unité de R^p, est une fonction croissante de r.

Le nombre $\nu(x) = \lim_{r=o} \nu(x,r)$ (nombre de Lelong du courant t au point x) joue le rôle d'une densité ; si $f = \log |F|$, F holomorphe, $\nu(x)$ est la multiplicité de F = 0 en x. Cette définition du nombre $\nu(x)$ n'est plus possible quand on passe à la dimension infinie. Pour y remédier, dans [1,b] nous avons construit $\nu(x)$ directement à partir d'une fonction plurisousharmonque f solution au voisinage du point x de

$$(1) \qquad t = \frac{i}{\pi}d'd''f.$$

Cette méthode conduit à des applications nouvelles qu'on donne ici succinctement.

2. On supposera les espaces vectoriels séparés, localement convexes, séquentiellement complets, $P(G)$ le cône des fonctions plurisousharmoniques dans un domaine G de E.

PROPOSITION 1. - *Soit* G *un domaine (ouvert connexe) d'un espace vectoriel topologique complexe* E *et* $f \in P(G)$. *Soit* $x_o \in G$; *il existe une suite croissante* $G_m \subset G_{m+1}$ *de domaines* $G_m \subset G$, *contenant* x_o *tel que* $G = \bigcup_m G_m$ *et que* f *soit bornée supérieurement dans* G_m.

La propriété est vraie de toute fonction semi-continue supérieurement : deux points $x_o \in G$, $y \in G$, appartiennent en effet à un compact connexe $K \subset G$, constitué par un *nombre fini* de segments réels et l'ensemble $[x \in G \ ; \ f(x) < m]$ pour $m > \sup_{x \in K} f(x)$ a donc une composante ouverte $G_m \subset G$ qui contient x_o et y ; on a ainsi $G = \bigcup G_m$ pour m parcourant les entiers positifs.

THÉORÈME 1. - *Soit* G *un domaine dans un espace vectoriel séparé et* $f \in P(G)$: $\nu(x)$ *relatif au courant positif* $\frac{i}{\pi} d'd''f$ *a la propriété suivante* :

Dans tout domaine G'_m *strictement intérieur à* G_m, $-\nu(x)$ *est l'enveloppe supérieure d'une famille de fonctions plurisousharmoniques négatives*.

Par domaine G'_m strictement intérieur à un domaine G_m nous entendons un ouvert connexe pour lequel il existe un voisinage ouvert W de l'origine tel qu'on ait

$$G'_m + W \subset G_m .$$

Démonstration. Pour $x \in G'_m$, $y \in W$, $r < 1$, on définit

$$(1) \quad -\nu(x,y,r) = (\log \frac{1}{r})^{-1} \left[\frac{1}{2\pi} \int_0^{2\pi} f(x+rye^{i\theta})d\theta - M_m \right]$$

où $M_m = \sup f(x)$ pour $x \in G_m$; le second membre de (1) est pour $r < 1$ une fonction négative, plurisousharmonique de $(x \times y) \in G'_m \times W$; elle est décroissante de r. Pour x fixé, la fonction

(2) $$- \nu(x,y) = \lim_{r=0} \left[-\nu(x,y,r)\right]$$

est définie pour tout $y \in E$ et vérifie $-\nu(x, \lambda y) = -\nu(x,y)$ pour tout $\lambda \in \mathbb{C}$. Ainsi la régularisée supérieure

$$\operatorname{Reg} \sup_y \left[-\nu(x,y)\right]$$

est indépendante de y ; soit $\alpha(x)$ sa valeur. On a donc

(3) $$-\nu(x,y,r) \leqslant -\nu(x,y) \leqslant \alpha(x)$$

et $-\nu(x,y) = \alpha(x)$ sauf pour des y appartenant à un cone polaire γ_x dans E.

Appliqué en dimension finie, le même calcul donne

$$\alpha(x) = -\nu(x)$$

et nous l'utiliserons ici pour définir $\nu(x)$. On a alors d'après (1) et (2)

$$-\nu(x) = \sup_{y,\,r}\left[-\nu(x,y,r)\right] \quad \text{pour } y \in W \,,\; r < 1$$

ce qui établit l'énoncé. On en déduit

THÉORÈME 2. - Soit M un sous-espace analytique de codimension un, analytiquement connexe dans un domaine G d'un espace localement convexe séparé E séquentiellement complet : la restriction de $-\nu(x)$ à M a une régularisée supérieure constante $-C_M$; l'ensemble $\nu(x) \geqslant C'$, $C' > C_M$ est localement polaire sur la variété connexe $\overset{\circ}{M}$, ensemble des points ordinaires de M.

La démonstration de ce second énoncé, à partir du théorème 1 suit fidèlement celle que nous avons donnée en dimension finie, cf. [1,c] et [1,d]. Elle utilise notamment l'énoncé suivant (avec la même démonstration que dans [1,d]) :

THÉORÈME 3. - Soit G un domaine dans E espace localement convexe comme au théorème 2, et soit Δ le domaine du produit $G \times C^n$ défini par $x \in G$, $0 < \|z\| < a$. Alors si $f(x,z)$ est plurisousharmonique dans Δ, et si l'on pose $L(x,r) = \sup f(x,z)$ pour x donné, et $\|z\| = r$, la fonction $$c(x) = \lim_{r=0} \left(\log \frac{1}{r}\right)^{-1} L(x,r)$$

a la propriété suivante : il existe c_o, $-\infty < c_o \leq +\infty$ tel qu'on ait $c(x) \leq c_o$ pour tout $x \in G$ et que l'ensemble $c(x) \leq c < c_o$ soit polaire dans tout domaine $G_m \subset G$ où $L(x,\frac{a}{2})$ est borné supérieurement.

3. Le théorème 3 montre encore ceci : dans E la singularité "la plus petite" pour les fonctions plurisousharmoniques est constituée par la classe des sous-ensembles analytiques de codimension 1 pour lesquels les points ordinaires constituent une variété connexe $\mathring{M}$. Encore faut-il pour qu'un tel ensemble M soit réellement une singularité que $\mathring{M}$ ne contienne pas un ensemble A non localement polaire sur $\mathring{M}$ aux points x desquels on a

$$\liminf_{x' \to x} f(x') \left[\log \frac{1}{p(x'-x)}\right]^{-1} = 0$$

par rapport à une semi-norme p(x) de E , quand $x' \to x$ selon une même direction indépendante de $x \in A$. Si E est un espace de Banach, on voit qu'on doit avoir $\liminf f(x) \left[\log d_M^{-1}(x)\right]^{-1} > 0$ où d_M est la distance de x à M. En particulier un théorème récent de Y.T.SIU [2] est valable en dimension infinie : si $\mathring{M}$ contient un point x, ayant un voisinage U(x) dans lequel f demeure borné supérieurement, M n'est pas une singularité : cet énoncé appliqué aux fonctions $\log |F|$, F holomorphe, généralise aussi aux espaces vectoriels topologiques un résultat classique de P.THULLEN.

4. De (3) résulte encore la possibilité de calculer $\nu(x)$ à partir d'une famille de sous-espaces de dimension finie convenablement choisie :

THÉORÈME 4. - La mesure positive $\nu(x)$ au point x peut être calculée à partir d'une classe de sous-espaces affines de dimension finie contenant x . Soit N un tel sous-espace de dimension n qui n'est pas contenu dans le cône polaire γ_x qui apparait dans la démonstration du théorème 1 et est défini comme l'ensemble des $y \in E - \{0\}$ pour lesquels on a $\nu(x,y) > \nu(x)$. Alors si f_N est la restriction de la fonction plurisousharmonique à N, et si $\sigma(x,r)$ est la mesure laplacien $\frac{1}{2\pi}\Delta f_N$ portée par la boule de centre x, de rayon r (pour une norme sur N) le quotient

$$\nu_N(x,r) = \left[\tau_{2n-2}\ r^{2n-2}\right]^{-1} \sigma(x,r)$$

est une fonction croissante de r et l'on a

$$\nu(x) = \lim_{r=o} \nu_N(x,r)$$

quel que soit le sous-espace N de dimension complexe n satisfaisant à la condition indiquée.

B I B L I O G R A P H I E

1. LELONG (P.). - a/ Intégration sur un ensemble analytique complexe. Bull. Soc. Math. de France, t. 85, p. 239-262, 1957.

   b/ Plurisubharmonic functions in topological vector spaces : polar sets and problems of measure. Proc. on infinite dimensional holomorphy. Lecture Notes Springer, n° 364, 1974.

   c/ Sur la structure des courants positifs fermés. C.R.Ac.Sci., t. 263, série A, p. 449-452, 1976.

   d/ Sur la structure des courants positifs fermés. Séminaire d'Analyse P.Lelong, Lecture Notes Springer, n° 578, p. 136-156, 1977.

2. SIU (Y.T.). - Analyticity of sets associated Lelong numbers and the extension of closed positive currents. Inv. Math., t. 27, p. 53-156, 1974.

A MAXIMAL EQUALITY AND ITS APPLICATION IN VECTOR SPACES

Raoul LePage
Department of Statistics and Probability
Michigan State University
East Lansing, Michigan 48824/USA

We will be concerned with a refinement of the method pioneered by J. Ville [3] whereby properties of a stochastic sequence with joint distribution P are derived by application of the maximal inequality to the sequence $\frac{dQ|F_n}{dP|F_n}$, $n \geq 0$ of Radon-Nikodym derivatives of an appropriately chosen positive σ-finite measure Q, taken over increasing σ-algebras $F_0 \subset F_1 \subset \ldots$ ad infinitum. This refinement will consist of a maximal equality, related to the maximal inequality, which may be used to closely approximate probabilities which the maximal inequality bounds from above.

It will be shown how limit theorems concerning the tail behavior of stochastic sequences may be established by such means. The case of stochastic sequences taking values in infinite dimensional spaces poses several problems for this method, some of which will be described here.

1. NOTATION. We use P to denote a probability measure on a σ-algebra F generated by the union of increasing σ-algebras $F_0 \subset F_1 \subset \ldots$ ad infinitum. The symbol Q will always denote a σ-finite positive measure absolutely continuous with respect to P on the algebra $\cup_n F_n$. From section 4 onward we will impose the condition $P \perp Q$ on F.

2. RADON-NIKODYM DERIVATIVES. For each $n \geq 0$ there is defined the Radon-Nikodym derivative $X_n = \frac{dQ|F_n}{dP|F_n}$, while on F we have

$$Q(A) = \int_{A(f>0)} X \, dP + Q(A(f = 0)) \qquad \forall A \in F$$

where $f = \frac{dP}{d(P+Q)}$ and $X = \frac{1-f}{f}$ on $(f > 0)$. We observe two simple facts for later reference. First, $X = \lim_{n\to\infty} X_n$ a.e. P and if $P \perp Q$ then $Q(\underline{\lim}_{n\to\infty} X_n < +\infty) = 0$. Second, $\{X_n, F_n, n \geq 0\}$ is an integrable P-martingale conditional on F_0, a.e. P. In essence, the assumed σ-finiteness of Q on F_0 means that Q behaves like a finite measure once F_0 is observed.

3. MAXIMAL INEQUALITY. We will use E, E^{F_n} to denote P-integration and conditional integration respectively, $\forall n \geq 0$. Also, events will, when placed under the integral, denote their indicator functions. For example, if λ is a positive

number, $E(X_0 \geq \lambda)$ will mean $P(X_0 \geq \lambda)$, whereas $EX_0(X_0 \geq \lambda)$ will mean $\int_{(X_0 \geq \lambda)} X_0 dP$, which is equal to $Q(X_0 \geq \lambda)$.

The following maximal inequality is then true,

$$(1) \qquad P(\exists n \geq 0,\ X_n \geq \lambda) \leq P(X_0 \geq \lambda) + \lambda^{-1} EX_0(X_0 < \lambda) \ .$$

Let's examine a proof of this inequality, in order to see why there is not equality. The proof we will follow is an extension of one given by Robbins ([1], line (15)) for a somewhat special case of this result.

On the event $(X_n < \lambda, \forall n \geq 0)$ define $\tau = +\infty$, elsewhere defining $\tau = \min\{n \geq 0\colon X_n \geq \lambda\}$.

Proof of (1): Since $P(\exists n \geq 0, X_n \geq \lambda) = P(\tau < \infty) = P(X_0 \geq \lambda) + P(0 < \tau < \infty)$, it is enough to bound $P(0 < \tau < \infty)$.

$$\begin{aligned} P(0 < \tau < \infty) &= \Sigma_1^\infty P(\tau = n) \\ &= \Sigma_1^\infty E(\tau \geq n) E^{F_{n-1}}(X_n \geq \lambda) \\ &\leq \lambda^{-1} \Sigma_1^\infty E(\tau \geq n) E^{F_{n-1}} X_n (X_n \geq \lambda) \qquad \text{(OVERSHOOT)} \\ &= \lambda^{-1} \Sigma_1^\infty EX_n(\tau = n) \ . \end{aligned}$$

In the above we have used the set inclusion $(\tau = n) \subset (X_n \geq \lambda)$ for a conditional form of the Markov inequality. Elsewhere we used $(\tau \geq n) \in F_{n-1}$ $\forall n \geq 1$. Continuing,

$$\begin{aligned} &= \lambda^{-1} \Sigma_1^\infty Q(\tau = n) \\ &= \lambda^{-1} Q(0 < \tau < \infty) \\ &\leq \lambda^{-1} Q(0 < \tau) \qquad \text{(NONSINGULARITY)} \\ &= \lambda^{-1} EX_0(X_0 < \lambda) \qquad \square \end{aligned}$$

Only two steps in this proof require inequalities. The first has been labelled "OVERSHOOT" since there is equality at this step provided $P(\tau < \infty, X_\tau > \lambda) = 0$. The second such step has been labelled "NONSINGULARITY" since there is equality at this step provided $P \perp Q$ on F.

4. TOWARD A MAXIMAL EQUALITY. This section is intended to motivate the maximal equality which will be proved in section 5, and is not required for the sequel. For this and all following sections assume $P \perp Q$ on F, where P, Q, F, F_n, $n \geq 0$ are as in section 1. From the discussion of the previous section we see that only overshoot of the boundary prevents equality in (1). To more closely examine the overshoot, we propose to smooth out the information $F_0, F_1, \ldots$, which

can be thought of as bursting in at times $0,1,\ldots$. What we will do is reveal this information more slowly over a continuous time scale.

For each $n \geq 1$, $r \geq 0$ define σ-algebras $F_{n,r} = \sigma\{F_{n-1},\{(X_n > s), s \leq r\}\}$. It is like moving out from 0 to ∞ and continuously asking "have I reached X_n yet?". Each σ-algebra $F_{n,r}$ contains the information F_{n-1} together with the knowledge as to whether $X_n \leq r$ and, if so, what the actual value of X_n is. It is not difficult to prove that, conditional on F_0, the function X_n is finitely integrable and the function $X_{n,r}$ defined by $X_{n,r} = \frac{dQ|F_{n,r}}{dP|F_{n,r}}$ satisfies $\forall n \geq 1$

$$X_{n,r} = \begin{cases} X_n & \text{on } (X_n \leq r) \\ E^{F_{n-1}}(X_n|X_n > r) & \text{on } (X_n > r) \end{cases}$$

where $E^{F_{n-1}}(X_n|X_n > r) \overset{\text{def.}}{=} \dfrac{E^{F_{n-1}}X_n(X_n > r)}{E^{F_{n-1}}(X_n > r)}$.

It will be proved in section 5 that $E^{F_{n-1}}(X_n|X_n > r)$ is defined for $r < X_n$, exceeds r, and is a.s. nondecreasing in r, $\forall n \geq 0$. The following pictures give some idea of how $X_{\cdot,\cdot}$ sample paths behave.

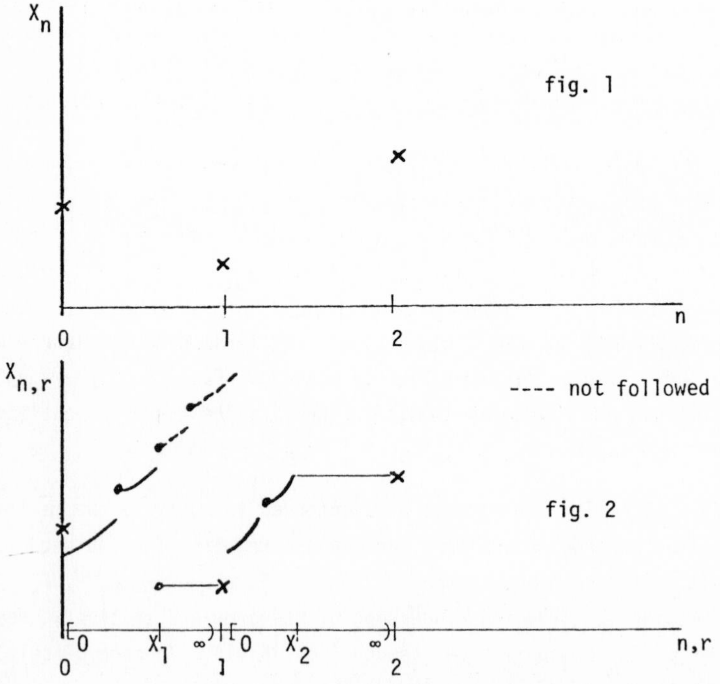

fig. 1

fig. 2

In figure 2, a possible graph of $X_{\cdot,\cdot}$ has been superimposed on figure 1. For each $n \geq 1$, except for the fact that $(X_{n-1} = 0) \subset (X_{n,0} = 0)$ a.e. P, there is no necessary size relationship between X_{n-1} and $X_{n,0}$. The central facts are that, for $n \geq 1$, $X_{n,r}$ increases in r, discontinuously at $\mathcal{F}_{n-1}$ conditional P-atoms of X_n, until it reaches $r = X_n$. Upon reaching $r = X_n$, the path of $X_{n,r}$ departs from the curve $E^{\mathcal{F}_{n-1}}(X_n | X_n > r)$, never discontinuously increasing at this point, thereafter remaining constant at $X_{n,r} = X_n$ for $r \geq X_n$.

Notice that by interposing the fields $\mathcal{F}_{\cdot,\cdot}$ we have eliminated any overshoot not occurring at precisely placed atoms. Notice also, for each $n \geq 1$ and $\lambda > 0$, that $\exists r \geq 0$ with $X_{n,r} \geq \lambda$ if and only if the left limit of the curve $E^{\mathcal{F}_{n-1}}(X_n | X_n > r)$ at $r = X_n$, namely $E^{\mathcal{F}_{n-1}}(X_n | X_n \geq r)|_{r-X_n}$, is $\geq \lambda$. This suggests that, in the absence of atoms in the conditional P-distributions, in which case $X_0 = X_{1,0}$ a.e. P, it should be true when $P \perp Q$ on $\mathcal{F}$ that,

$$
(2) \qquad P(\exists n \geq 1, E^{\mathcal{F}_{n-1}}(X_n | X_n \geq r)|_{r=X_n} \geq \lambda)
$$
$$
= P(X_0 \geq \lambda) + \lambda^{-1} E X_0(X_0 < \lambda).
$$

This result is true, and is a special case of the result to be proved in section 5.

We turn now to the problem of smoothing atoms. Suppose, for some $n \geq 1$, $\lambda > 0$, and $r' \geq 0$, that r' is an $\mathcal{F}_{n-1}$-conditional P-atom in the distribution of X_n, and that this could cause overshoot. This situation is characterized by $P^{\mathcal{F}_{n-1}}(X_n > r') > 0$ and the inequalities $m_n(r'-) < \lambda \leq m_n(r')$, where $m_n(r'-) \overset{\text{def.}}{=} E^{\mathcal{F}_{n-1}}(X_n | X_n \geq r')$ and $m_n(r') \overset{\text{def.}}{=} E^{\mathcal{F}_{n-1}}(X_n | X_n > r')$. In such a case, overshoot will occur if and only if $X_n > r'$. No overshoot occurs if $X_n = r'$ since, as remarked earlier, there is no possibility of discontinuous increase by $X_{n,r}$ at $r = X_n$. To smoothly reveal whether or not $X_n > r'$ requires independent randomization.. Suppose there is an $\mathcal{F}_{n,r'}$-measurable function $\mathcal{U}$ whose distribution with respect to P, and also Q, is Lebesgue measure on $[0,1]$. That is, $\mathcal{U}$ is a random number, independent of $\mathcal{F}_{n,r'-}$, with respect to each of P and Q. The existence of such a $\mathcal{U}$ is not a restrictive assumption since one can enlarge the space by taking the product with a unit interval endowed with Lebesuge measure. Define Y as follows:

$$
Y = Y_{n,r'} \overset{\text{def.}}{=} \begin{cases} 1 & \text{on } (X_n > r') \\ \mathcal{U} & \text{on } (X_n = r') . \end{cases}
$$

Think now of successively asking the questions "is $Y > u$?" as u moves from 0 toward 1. So long as the answer remains "yes" it is still possible that $X_n > r'$.

However, if the asnwer for some $u < 1$ is "no", then $X_n = r'$ is assured. Conditional on F_0, and on the event $(X_n \geq r')$, it is not difficult to prove that the appropriately defined $X_{n,r'-,u}$, $0 \leq u < 1$, satisfy:

$$X_{n,r'-,u} = \begin{cases} \dfrac{r'\pi(1-u) + (1-\pi)m}{\pi(1-u) + (1-\pi)} & \text{on } (u < Y) \\ X_n & \text{on } (u \geq Y) \end{cases}$$

where

$$\pi = \pi_n(r') \stackrel{\text{def.}}{=} E^{F_{n-1}}(X_n = r' | X_n \geq r')$$

$$m = m_n(r') \stackrel{\text{def.}}{=} E^{F_{n-1}}(X_n | X_n > r') \qquad \text{on } (P^{F_{n-1}}(X_n > r)|_{r=X_n} > 0).$$

Once again, there can be no increasing discontinuity at the point where $X_{n,r'-,u}$ departs from a nondecreasing curve to take the value X_n. Here is a possible path:

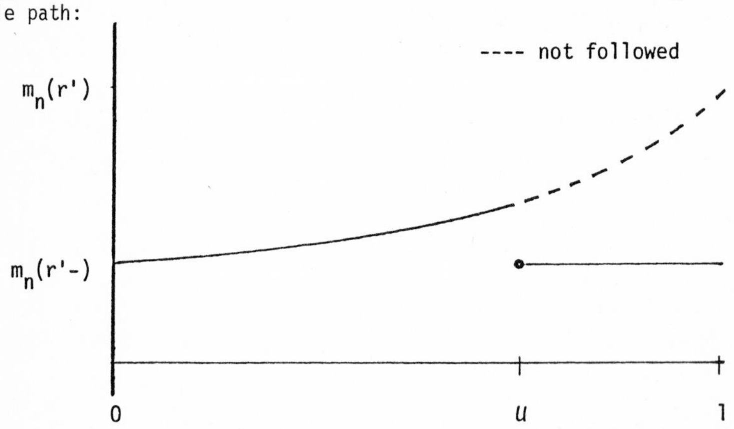

fig. 3

Notice that $\exists u \in [0,1)$ with $X_{n,r'-,u} \geq \lambda$ if and only if $X_n > r'$ (i.e. $E^{F_{n-1}}(X_n|X_n \geq r') \geq \lambda$), or $X_{n+1} = r'$ (i.e. $E^{F_n}(X_{n+1} > r') \geq \lambda$) and $X_{n,r'-,u}|_{u=U} \geq \lambda$. The latter event has probability $p \stackrel{\text{def.}}{=} 1-u'$ where $\dfrac{r'\pi(1-u') + (1-\pi)m}{\pi(1-u') + (1-\pi)} = \lambda$. The solution to the previous equation is, $p = \dfrac{m-\lambda}{\lambda-r'} \dfrac{1-\pi}{\pi}$, which can be also expressed as $p = \dfrac{m_n(X_n) - \lambda}{\lambda - X_n} \dfrac{1 - \pi_n(X_n)}{\pi_n(X_n)}$. In section 5 it will be proved this is a.s. p a number in $[0,1]$. It seems therefore that the following maximal equality, generalizing (2), should be valid for every $\lambda > 0$, when $P \perp Q$ on F,

(2') $\quad P(X_0 \geq \lambda$, or $\exists n \geq 0$ with $m_n(X_n-) \geq \lambda$, or $\exists n \geq 0$ with $m_n(X_n-) < \lambda \leq m_n(X_n)$ and an event with conditional probability p (given above) occurs)

$$= P(X_0 \geq \lambda) + \lambda^{-1} EX_0(X_0 < \lambda).$$

This is correct, and will be proved in the next section.

Although the statement of (2') is somewhat extensive, the idea is a simple one. Instead of looking to stop at the first $n \geq 0$ at which $X_n \geq \lambda$, look for $X_0 \geq \lambda$, and (for $n \geq 1$) for $m_n(X_n-) \geq \lambda$, where $m_n(X_n-) \overset{\text{def.}}{=} E^{F_{n-1}}(X_n | X_n \geq r)|_{r=X_n} \geq X_n$. If for some n, the latter is $< \lambda$, but the still larger $m_n(X_n)$ $\overset{\text{def.}}{=} E^{F_{n-1}}(X_n|X_n > r)|_{r=X_n}$ is $\geq \lambda$ (indicating an unfortunately placed atom) then randomly stop with the conditionally defined probability p. With this modification, the P-probability of ever stopping is (for $P \perp Q$ on F) exactly given by the right side of (2').

5. MAXIMAL EQUALITY. We begin with a result which abstracts the discoveries of the previous section.

For a random variable X with expectation $\mu < +\infty$, let

$$\gamma = \sup\{r: P(X > r) > 0\}$$

$$m(r) = \begin{cases} \int_{X>r} XdP/P(X > r) & r < \gamma \\ \gamma & r \geq \gamma \end{cases}$$

In the above definitions, μ may be $-\infty$, r is free to range over $(-\infty,\infty)$, and γ may be $+\infty$ (in which case the set $r \geq \gamma$ is void).

The function m, associates to each $r < \gamma$ the conditional mean of X on the event $(X > r)$, for which we will write $E(X|X > r)$. It is seen to be nondecreasing and continuous from the right. Additional properties of interest (also immediate) are: $\lim_{r\to-\infty} m(r) = \mu$, $\lim_{r\to\infty} m(r) = \gamma$, and $r < m(r)$ for $r < \gamma$. The preceeding remarks imply that $X \vee \mu \leq m(X) \leq \gamma$.

Proposition 1. If X is a nondegenerate random variable with expectation $\mu < +\infty$, then $\mu < m(\gamma-)$, $P(m(X-) \geq \lambda) > 0$ for every $\lambda \in (\mu, m(\gamma-))$, and

(3) $$E(X|m(X) \geq \lambda) \leq \lambda$$

(4) $$E(X|m(X-) \geq \lambda) \geq \lambda .$$

If $\mu > -\infty$ then (4) is correct for $\lambda = \mu$. If $m(\gamma-) < \infty$ (equivalent to $\gamma < \infty$)

then $m(\gamma-) = \gamma$ and (4) is correct for $\lambda = \gamma$.

Proof. The proof for the cases $\mu > -\infty$, $\lambda = \mu$ and $\gamma < \infty$, $\lambda = \gamma$ is easy, and will be omitted. If $\lambda < m(\gamma-)$ there exist $r < \gamma$ with $m(r) \geq \lambda$. For such r, $0 < P(X > r) \leq P(m(X-) \geq \lambda)$ by monotonocity of m. Let λ be any point of $(\mu, m(\gamma-))$. To prove (3) let $r_\lambda = \min\{r: m(r) \geq \lambda\}$, which exists by right continuity of m. Clearly $m(X) \geq \lambda$ if and only if $X \geq r_\lambda$. The conditional mean appearing on the left side of (3) is therefore $m(r_\lambda-)$, which does not exceed λ. To prove (4) first consider the case $m(r_\lambda-) = \lambda$. In such a case, $m(X-) \geq \lambda$ if and only if $X \geq r_\lambda$, from which it follows that the left side of (4) is equal to $m(r_\lambda-) = \lambda$. If, on the other hand, $m(r_\lambda-) < \lambda$, then $m(X-) \geq \lambda$ if and only if $X > r_\lambda$. In this second case, the left side of (4) is equal to $m(r_\lambda) \geq \lambda$. □

To paraphrase part of the preceeding proposition, if a person is selected at random and we are told that the average height of persons at least as tall as this person is at least 6 feet, then our conditional expectation of this person's height is nearly 6 feet. In fact, as shown by the following corollary, our conditional expectation of the person's height is exactly 6 feet provided we assume heights of people are continuously distributed.

Corollary 1. If $\lambda \in (\mu, m(\gamma-))$, and $P(X = r_\lambda) = 0$ (equivalent to r_λ being a continuity point of m), or if $m(r_\lambda-) = \lambda$, then (3) and (4) are equalities, and are the same.

Proof. If $P(X = r_\lambda) = 0$ the result follows immediately from Proposition 1. The case $m(r_\lambda-) = \lambda$ is familiar form the proof of Proposition 1. It was proved that $m(X-) \geq \lambda) = (X \geq r_\lambda)$ in this case, and that $(X \geq r_\lambda) = (m(X) \geq \lambda)$ in general.□

The next result shows how an independent randomization can achieve equality when Corollary 1 is inapplicable.

Corollary 2. If $\lambda \in (\mu, m(\gamma-))$, and if $P(X = r_\lambda) > 0$ and $m(r_\lambda-) < \lambda$, then $\pi = P(X = r_\lambda)/P(X \geq r_\lambda)$ defines a number in $(0,1)$ and the equation

$$\text{(5)} \qquad \frac{\pi p r + (1 - \pi)\, m(r_\lambda)}{\pi p + (1 - \pi)} = \lambda$$

has the unique solution $p_\lambda = \dfrac{m(r_\lambda) - \lambda}{\lambda - r_\lambda} \dfrac{1 - \pi}{\pi} \in (0,1)$. Let H be an event independent of X, with $P(H) = P_\lambda$ as defined below (5). We will call H an auxilliary event. The left side of (5) is then equal to the conditional mean of X on the event $A = (X > r_\lambda) \cup ((X - r_\lambda) \cap H)$.

Proof. The last part of the corollary is self evident. We first prove the assertion concerning π. The proof of Proposition 1 has shown that $(m(X) \geq \lambda) = (X \geq r_\lambda)$ in general, while $(m(X-) \geq \lambda) = (X > r_\lambda)$ in the particular case $m(r_\lambda-) < \lambda$ assumed here. Therefore $P(X > r_\lambda) = P(m(X-) \geq \lambda)$, which is positive by Proposition 1. This proves $\pi \in (0,1)$. To prove $p_\lambda \geq 0$, notice first that $\lambda \leq m(r_\lambda)$, so it is enough to prove $r_\lambda < \lambda$. This follows from the inequalities

$\lambda > m(r_\lambda -) = \pi r_\lambda + (1-\pi)m(r_\lambda) \geq \pi r_\lambda + (1-\pi)\lambda$ and $\pi > 0$. Therefore p_λ is defined and is nonnegative. It is easy to check that p_λ solves (5). The only part remaining is $p_\lambda < 1$, but this is algebraically equivalent to $\pi r_\lambda + (1-\pi)m(r_\lambda) < \lambda$, which was established above. □

In the course of proving Corollary 2, it was established that the event A has the alternative description $A = (m(X-) \geq \lambda) \cup ((m(X-) < \lambda \leq m(X)) \cap H)$.

For each $n \geq 1$ define $m_n(X_n -) \overset{\text{def.}}{=} E^{F_{n-1}}(X_n | X_n \geq r)|_{r=X_n}$, and on the event $(P^{F_{n-1}}(X_n > r)|_{r=X_n} > 0)$ define $m_n(X_n) = E^{F_{n-1}}(X_n | X_n > r)|_{r=X_n}$. Also, for each $n \geq 1$, and $\lambda > 0$, define the events $A_n = A_{n,\lambda} \overset{\text{def.}}{=} (m_n(X_n -) \geq \lambda$, or $m_n(X_n -) < \lambda \leq m_n(X_n)$ and $H)$ where H is an auxilliary event of conditional probability $p = \dfrac{m_n(X_n) - \lambda}{\lambda - X_n} \dfrac{1 - \pi_n(X_n)}{\pi_n(X_n)}$ and $\pi_n(X_n) = P^{F_{n-1}}(X_n = r | X_n \geq r)|_{r=X_n}$. Define the (improper) stopping random variable τ_1 as follows:

$$\tau_1 \overset{\text{def.}}{=} \begin{cases} 0 & \text{on } X_0 \geq \lambda \\ n & \text{on } (X_0 < \lambda) \bigcap_1^{n-1} A_k^c \cap A_n, \ \forall n \geq 1 \\ +\infty & \text{on } \bigcap_1^{\infty} A_k^c \end{cases}$$

Proposition 2 (maximal equality). If $P, Q, F, F_n, X_n, n \geq 0$ are as in sections 1 and 2, and if $P \perp Q$ on F, then for every $\lambda > 0$

$$(2') \qquad P(\tau_1 < \infty) = P(X_0 \geq \lambda) + \lambda^{-1} EX_0(X_0 < \lambda).$$

Proof. $P(\tau_1 < \infty) = P(X_0 \geq \lambda) + P(0 < \tau_1 < \infty)$. We will apply Corollary 2 in its conditional form.

$$\begin{aligned} P(0 < \tau_1 < \infty) &= \sum_1^\infty P(\tau_1 = n) \\ &= \sum_1^\infty E(\tau_1 \geq n) E^{F_{n-1}} A_n \\ &= \sum_1^\infty E(\tau_1 \geq n) \lambda^{-1} E^{F_{n-1}} X_n A_n & \text{(Corollary 2)} \\ &= \lambda^{-1} \sum_1^\infty Q(\tau_1 = n) \\ &= \lambda^{-1} Q(0 < \tau_1 < \infty) \\ &= \lambda^{-1} Q(0 < \tau_1) & (\tau_1 \leq \tau \text{ and } P \perp Q) \\ &= \lambda^{-1} Q(X_0 < \lambda) \\ &= \lambda^{-1} EX_0(X_0 < \lambda). \end{aligned}$$

□

The following example suggests how (6) may be used to study the tail behavior of the sequence of successive maxima of uniform random variables.

6. APPLICATION TO SUCCESSIVE MAXIMA. For each $\theta > 0$, denote by P_θ the probability distribution on (R^∞, B^∞) with respect to which the coordinate random variables ξ_n, $n \geq 1$ are independent and identically distributed with the uniform distribution on the interval $[0,\theta]$. Define $P \overset{\text{def.}}{=} P_1$, $Q \overset{\text{def.}}{=} \int_{0+}^{1} P_\theta d\theta$, and denote the successive maxima by $\xi_{(n)} \overset{\text{def.}}{=} \max\{\xi_1,\ldots,\xi_n\}$ $\forall n \geq 1$. Only a short calculation is required to show that, with $F_n \overset{\text{def.}}{=} \sigma\{\xi_1,\ldots,\xi_n\}$, the derivatives of Q by P

are $X_n = \frac{1}{n-1}((\frac{1}{\xi_{(n)}})^{n-1} - 1)$ $\forall n > 1$. Since $\xi_{(n)} \geq \xi_{(n-1)}$, it is obvious that $m_n(X_n) < \frac{1}{n-1}((\frac{1}{\xi_{(n-1)}})^{n-1} - 1)$ $\forall n > 1$. If $\lambda > 0$ and j is a positive integer, $F_n \overset{\text{def.}}{=} \sigma\{\xi_k, k \leq n+j\}$ $\forall n \geq 0$. Then from (2') and the preceeding remark, $\forall j > 1$,

$$\text{(6)} \qquad P(\xi_{(n)} \leq (1 + (n-1)\lambda)^{-\frac{1}{n-1}} \text{ for some } n \geq j)$$

$$\leq P(\xi_{(j)} \leq (1 + (j-1)\lambda)^{-\frac{1}{j-1}}) + \int_{0+}^{1} P(\xi_{(j)} > \theta^{-1}(1 + (j-1)\lambda)^{-\frac{1}{j-1}})d\theta$$

$$\leq P(\xi_{(n)} \leq (1 + n\lambda)^{-\frac{1}{n}} \text{ for some } n \geq j).$$

That is, the middle term of (6) is an upper bound for the probability with which $\xi_{(n)}$, $n \geq j$ ever downwardly crosses the boundary $(1 + (n-1)\lambda)^{-1/n}$, $n \geq j$. It is also a lower bound for the same probability when the boundary is shifted by one time unit. Making use of the fact that $(1 + n\lambda)^{1/n} \geq (1 + (n-1)\lambda h)^{1/n-1}$ $\forall n \geq j > 1$, where $h = (j-1)^{-1}\,\lambda^{-1}((1 + j\lambda)^{j-1/j} - 1)$, one may obtain the following from (6), by taking $\lambda = \frac{\alpha}{j}$ and letting $j \to \infty$:

$$\text{(7)} \qquad \lim_{j\to\infty} P(\xi_{(n)} \leq (1 + \frac{n\alpha}{j})^{-\frac{1}{n}} \text{ for some } n \geq j) = \alpha^{-1} \log_e(1 + \alpha), \ \forall \alpha > 0.$$

This says that for large j the sequence $(1 + \frac{n\alpha}{j})^{-\frac{1}{n}}$, $n \geq j$ is a lower bound for the sequence $\xi_{(n)}$, $n \geq j$ with probability almost exactly given by $\alpha^{-1}\log_e(1 + \alpha)$. Since the right side of (7) is nonincreasing between the limits 1 and 0 on $(0,\infty)$, one suspects there is a convergence in distribution. Indeed, from (7) follows

$$\text{(8)} \qquad \lim_{j\to\infty} P(j \sup_{n\geq j} n^{-1}(\xi_{(n)}^{-n} - 1) \geq \alpha) = \alpha^{-1}\log_e(1 + \alpha), \quad \forall \alpha > 0.$$

To get (8) from (7) simply use the fact that $P(\xi_{(n)} = (1 + \frac{n\alpha}{j})^{-\frac{1}{n}}$ for some $n \geq j) = 0$ for all $\alpha > 0$, $j > 1$.

The work of Ville, Robbins, and others developing Ville's method, has been largely restricted to real random variables. The deepest results so far obtained have been with Q taken to be an appropriately defined mixture of probability measures from an exponential family containing P. Unfortunately, the situation in infinite dimensional space is complicated by the fact that there is much singularity. Gaussian measures with infinite dimensional support, for example, are mutually singular with different scalings, a fact which effectively blocks some of the most fruitful constructions for Q (for examples, see [1], [3]). Even when there are interesting mutually absolutely continuous measures, such as the translations of a Gaussian measure by kernel elements, the set of these measures is too sparse to admit interesting σ-finite mixtures Q. The example to follow, which deals with the Gaussian case, suggests an approach which may have some useful consequences.

7. APPLICATION TO SUCCESSIVE AVERAGES OF A GAUSSIAN SEQUENCE. Suppose X_n, $n \geq 1$ are independent random variables distributed according to a zero mean Gaussian measure Γ on a separable Banach space B. Define $\overline{\mu_0^*} = 0 \in B^*$, and $\overline{X}_n = \frac{1}{n}(X_1 + \ldots + X_n)$ $\forall n \geq 1$. For $n \geq 1$ let $\overline{\mu_n^*} = \frac{1}{n}(\mu_1^* + \ldots + \mu_n^*)$ where μ_n^* is recursively defined to be the continuous linear function of minimum norm satisfying $\overline{\mu_n^*}(\overline{X}_n) = \|\overline{X}_n\|_B^2$ and $\mu_n^*(\overline{\mu}_{n-1}) = \overline{\mu_{n-1}^*}(X_n)$, where $\overline{\mu}_{n-1}$ is the kernel element corresponding to $\overline{\mu_{n-1}^*}$. The first of the equations just given is equivalent to $\mu_n^*(\overline{X}_n) = n\|\overline{X}_n\|_B^2 - (n-1)\overline{\mu_{n-1}^*}(\overline{X}_n)$, and since $\overline{X}_n$ is a.e. not colinear with $\overline{\mu}_{n-1}$ (unless the dimension of the kernal space is one, in which case the two equations are compatible) the two equations are compatible. Therefore μ_n^* and μ_n are well defined by the Hahn-Banach theorem, and may even be chosen measurably in $X_1, \ldots, X_n$, $\forall n \geq 1$ (examine the proof in [2], pg. 112). Define $F_n = \sigma\{X_k, k \leq n\}$ $\forall n \geq 1$, $F_0 = \{\phi, B\}$, P = probability distribution of X_n, $n \geq 1$, and define Q to be the probability measure characterized through its conditional distributions $Q^{F_n}|F_{n+1} = \Gamma(\cdot - \overline{\mu}_n)$ $\forall n \geq 0$. What we are trying to do is construct a Q which can compete with P for probability in the neighborhood of the observations X_n, $n \geq 0$ (without vitiating $P \perp Q$ on F, which would make the maximal inequality inaccurate), thus achieving sharp information from the maximal inequality. Then,

$$\frac{dQ|F_n}{dP|F_n} = e^{\sum_1^n \overline{\mu_{k-1}^*}(X_k) - \frac{1}{2}\sum_1^n \overline{\mu_{k-1}^*}(\overline{\mu}_{k-1})}, \ \forall n \geq 1.$$

For each $\lambda > 0$ the inequalities $\frac{dQ|F_n}{dP|F_n} < \lambda$, $\forall n \geq 1$ may be rewritten $Y_n < 2 \log_e \lambda$, $\forall n \geq 1$, where

$$Y_n = 2\sum_1^n \overline{\mu_{k-1}^*}(X_k) - \sum_1^n \overline{\mu_{k-1}^*}(\overline{\mu}_{k-1}) .$$

By elementary algebraic operations,

$$Y_n = \sum_1^n \mu_k^*(X_k) - \sum_1^n (\mu_k^* - \overline{\mu_{k-1}^*})(X_k - \bar{\mu}_{k-1}) \qquad \text{(definition of } \mu_k^*)$$

$$= n\|\bar{X}_n\|_B^2 + \sum_1^n (\mu_k^* - \overline{\mu_n^*})(X_k - \bar{\mu}_n) - \sum_1^n (\mu_k^* - \overline{\mu_{k-1}^*})(X_k - \bar{\mu}_{k-1})$$

$$= n\|\bar{X}_n\|_B^2 + Z_n.$$

To simplify the above notice that $\forall n \geq 1$,

$$Z_n = \sum_1^n (\mu_k^* - \overline{\mu_n^*})(X_k - \bar{\mu}_n) - \sum_1^n (\mu_k^* - \overline{\mu_{k-1}^*})(X_k - \bar{\mu}_{k-1})$$

$$= \sum_1^n (\mu_k^* - \overline{\mu_{n-1}^*})(X_k - \bar{\mu}_{n-1}) - \sum_1^n (\mu_k^* - \overline{\mu_{k-1}^*})(X_k - \bar{\mu}_{k-1}) - n(\overline{\mu_n^*} - \overline{\mu_{n-1}^*})(\bar{X}_n - \bar{\mu}_{n-1})$$

$$= Z_{n-1} - n(\overline{\mu_n^*} - \overline{\mu_{n-1}^*})(\bar{X}_n - \bar{\mu}_{n-1}).$$

From the above,

$$Y_n = n\|\bar{X}_n\|_B^2 - \sum_1^n k(\overline{\mu_k^*} - \overline{\mu_{k-1}^*})(\bar{X}_k - \bar{\mu}_{k-1}), \ \forall n \geq 1.$$

So, $\frac{dQ|F_n}{dP|F_n} < \lambda$, $\forall n \geq 1$ may be rewritten

$$\text{(9)} \qquad \|\bar{X}_n\|_B^2 < n^{-1}(2 \log_e \lambda + \sum_1^n k(\overline{\mu_k^*} - \overline{\mu_{k-1}^*})(\bar{X}_k - \bar{\mu}_{k-1})), \ \forall n \geq 1.$$

The maximal equality may be used to prove a limit theorem for the probability of the event in (9), much as was done in the previous section. While these results are incomplete, it can be pointed out that when B is finite dimensional and is the kernel space for Γ, (9) takes the form (with $\bar{X}_0$ defined to be zero),

$$\text{(10)} \qquad \|\bar{X}_n\|_B^2 < n^{-1}(2 \log_e \lambda + \sum_1^n k\|\bar{X}_k - \bar{X}_{k-1}\|_B^2) \qquad \forall n \geq 1 .$$

It is not difficult to show that the right side of (10) is $O(\frac{\log_e n}{n})$ a.e. Thus (9) may be viewed as an attempt to generalize (1) to infinite dimensions. Whether the right side of (9) is in general a.e. $O(\frac{\log_e n}{n})$ is still an open question. If it is, there is a role for the Hahn-Banach theorem in defining nonlinear functions of the data, taking values in the kernel space, which function acceptably as estimators of the mean.

References

[1] Robbins, H. (1970). Statistical methods related to the law of the iterated logarithm. Ann. Math. Statist. 41 1397-1409.

[2] Rudin, W. (1974). Real and Complex Analysis. Second edition. McGraw-Hill.

[3] Ville, J. (1939). Etudé Critique de la Notion de Collectif. Gauthier-Villars, Paris.

REPRESENTATION OF ANALYTIC FUNCTIONALS BY VECTOR MEASURES

by *Jorge Mujica*

INTRODUCTION. Let K be a compact locally connected set in the plane. A. Baernstein has shown in [2, p. 31] that given an analytic functional on K, i.e. a continuous linear functional T on $\mathcal{H}(K)$, the space of holomorphic germs on K, there exists a sequence (μ_m) of complex Borel measures on K such that

$$< f, T > = \sum_{m=0}^{\infty} \int_K \frac{1}{m!} f^{(m)}(x) d\mu_m(x)$$

for every $f \in \mathcal{H}(K)$. In this paper we extend Baernstein's result to the case where K is a compact locally connected subset of a complex metrizable Schwartz space E. In this extension, the complex measures μ_m are replaced by vector measures with values in the dual of $\mathcal{P}(^mE)$, the space of continuous m-homogeneous polynomials on E, and the derivatives $f^{(m)}(x) \in \mathbb{C}$ are replaced by the differentials $\hat{d}^m f(x) \in \mathcal{P}(^mE)$. This extension is given in Section 3, in Theorem 3.4. In Sections 1 and 2 we set up the necessary machinery. However, the main result in Section 1, namely Theorem 1.5, is also interesting in itself, and we give one further application of it in Section 4. A brief description of each section is the following.

In Section 1 we study the space $\mathcal{C}(X;F)$ of all continuous functions which are defined on a compact Hausdorff space X and with values in a compactly-regular (LB)-space $F = \text{ind lim } F_j$. In Theorem 1.5 we prove that $\mathcal{C}(X;F)$, endowed with the topology of uniform convergence on X, is topologically isomorphic to the inductive limit of the Banach spaces $\mathcal{C}(X;F_j)$.

In Section 2 we study certain spaces of sequences of continuous functions. These spaces of sequences are a generalization of those already considered by Baernstein in [2]. Given a compact Hausdorff space X and a sequence of compactly-regular (LB)-spaces $F_m = \text{ind lim } F_{mj}$, we define $\mathcal{S}_j$ to be the Banach space of all sequences $\Phi = (\phi_m)$ such that, for every m, $\phi_m \in \mathcal{C}(X;F_{mj})$ and

$$\|\Phi\|_j = \sup_m j^{-m} \|\phi_m\|_{mj} < \infty,$$

and define $\mathcal{S}$ to be the inductive limit of the Banach spaces $\mathcal{S}_j$. Then, using Theorem 1.5, we characterize the dual of $\mathcal{S}$ in terms of vector measures.

In Section 3 we give the announced extension of Baernstein's representation of $\mathcal{H}'(K)$. To get the desired result, we embed $\mathcal{H}(K)$ as a topological subspace of a suitable space of sequences of continuous functions and use the results of Section 2.

Finally, in Section 4 we consider the following question, raised by Bierstedt and Meise in [4]. Given a Banach space E and a compactly-regular (LB) - space $F = \text{ind lim } F_j$, is it true that the ε- product $E \varepsilon F$ is topologically isomorphic to the inductive limit of the ε-products $E \varepsilon F_j$? As an application of Theorem 1.5 we answer this question affirmatively in the case where $E = \mathcal{C}(X)$, X being a compact Hausdorff space.

I would like to thank J. B. Prolla for many useful discussions. I would also like to thank P. Berner and Ph. Boland for filling in a gap in the original proof of Theorem 3.1.

1. SPACES OF CONTINUOUS FUNCTIONS WITH VALUES IN AN INDUCTIVE LIMIT.

Throughout this section F denotes an (LB) - space, i.e. the inductive limit of an increasing sequence of Banach spaces F_j such that each inclusion mapping $F_j \hookrightarrow F_{j+1}$ is continuous and $F = \cup F_j$. Without loss of generality we may assume that

$$\| x \|_{j+1} \leqq \| x \|_j$$

for all $x \in F_j$. We devote this section to study some properties of the space of continuous functions $\mathcal{C}(X;F)$, where X denotes a compact Hausdorff space.

The usual topology on $\mathcal{C}(X;F)$ is the topology τ_u of uniform convergence on X, which is defined by the seminorms

$$f \in \mathcal{C}(X;F) \rightarrow \sup_{x \in K} q(f(x)) \in \mathbb{R}$$

where q varies among the continuous seminorms on F. Another natural topology on $\mathcal{C}(X;F)$ is the inductive topology τ_i with respect to the inclusion mappings $\mathcal{C}(X;F_j) \hookrightarrow \mathcal{C}(X;F)$. It is clear that $\tau_u \leqq \tau_i$. We will show that whenever $\mathcal{C}(X;F) = \cup \mathcal{C}(X;F_j)$, both topologies coincide.

First we will give sufficient conditions for $\mathcal{C}(X;F) = \cup \mathcal{C}(X;F_j)$ to happen. We recall the following definitions.

An inductive limit of subspaces $G = \text{ind lim } G_\alpha$ is said to be

(i) regular [7, p. 123] if each bounded subset of G is contained and bounded in some G_α ;

(ii) compactly - regular [3, p. 100] if each compact subset of G is contained and compact in some G_α ;

(iii) Cauchy - regular [9, Def. 1.5] if given a bounded subset B of G , there exists α such that B is contained and bounded in G_α , and furthermore, G and G_α induce the same Cauchy nets in B .

We remark that the term "Cauchy - regular" coincides, in the case of (LB) - spaces, with each of the terms "boundedly retractive" and "strongly boundedly retractive", introduced by Bierstedt and Meise in [3, p. 100].

The following result is clear.

1.1 PROPOSITION. If the inductive limit $F = \text{ind lim } F_j$ is compactly-regular, or in particular Cauchy - regular, then $\mathcal{C}(X;F) = \cup \, \mathcal{C}(X;F_j)$.

To prove that whenever $\mathcal{C}(X;F) = \cup \, \mathcal{C}(X;F_j)$ the topologies τ_u and τ_i coincide, we will show that both topologies yield the same dual and the same equicontinuous subsets in the dual. We will use the following result of Singer's [14 , p. 398, Th.], which characterizes the dual of $\mathcal{C}(X;G)$ in terms of vector measures, when G is a Banach space.

1.2 SINGER'S THEOREM. Let X be a compact Hausdorff space and let G be a Banach space. Then there is a one - to - one correspondence between the continuous linear functionals T on $\mathcal{C}(X;G)$ and the regular Borel measures μ on X , with values in G' , of bounded variation. This correspondence is given by the formula

$$<\phi , T> = \int_X \phi d\mu$$

for all $\phi \in \mathcal{C}(X;G)$, and the total variation $\|\mu\|$ of μ equals the norm $\|T\|$ of T .

1.3 LEMMA. If $\mathcal{C}(X;F) = \cup \, \mathcal{C}(X;F_j)$, then the topologies τ_u and τ_i on $\mathcal{C}(X;F)$ yield the same dual.

PROOF. Since $\tau_u \leqq \tau_i$, it is clear that $\mathcal{C}_u'(X;F) \subset \mathcal{C}_i'(X;F)$. We will show that the opposite inclusion also holds. Let $T \in \mathcal{C}_i'(X;F)$, let $T_j \in \mathcal{C}'(X;F_j)$ denote the restriction of T to $\mathcal{C}(X;F_j)$ and let $\mathcal{B}$ denote the σ - algebra of all Borel subsets of X. By Singer's Theorem 1.2, for each j there exists a unique set function $\mu_j : \mathcal{B} \to F_j'$ which is countably-additive, regular and of bounded variation, such that

$$< \phi , T_j > = \int_X \phi \, d\mu_j$$

for all $\phi \in \mathcal{C}(X;F_j)$. In view of the uniqueness of each μ_j, the following diagram is commutative whenever $j < k$.

$$\begin{array}{ccc} & \mathcal{B} & \\ {\scriptstyle \mu_k} \downarrow & \searrow {\scriptstyle \mu_j} & \\ F_k' & \longrightarrow & F_j' \end{array}$$

It follows that for each $B \in \mathcal{B}$ the sequence $(\mu_j(B))$ defines an element $\mu(B) \in \text{proj lim } F_j'$. Since F' is topologically isomorphic to proj lim F_j' for the weak topologies [12, Ch. IV, Prop. 4.5], it follows that $\mu : \mathcal{B} \to F'$ is a set function which is countably-additive and regular, and the following diagram is commutative for each j.

$$\begin{array}{ccc} & \mathcal{B} & \\ {\scriptstyle \mu} \downarrow & \searrow {\scriptstyle \mu_j} & \\ F' & \longrightarrow & F_j' \end{array}$$

We claim that the vector measure μ has total variation $\leqq 1$ with respect to a certain continuous seminorm q on F. To see this we will construct a convex, balanced 0-neighborhood W in F such that

$$(*) \qquad \sum_{i=1}^{n} | < w_i , \mu(B_i) > | \leqq 1$$

for each finite partition $\{B_1 , \ldots , B_n\}$ of X with $B_i \in \mathcal{B}$, and each $\{w_1 , \ldots , w_n\} \subset W$. After constructing such a W, the Minkowski functional q of W will do the job.

μ_j being of bounded variation, there exists a 0-ball V_j in F_j such that

$$(**) \qquad \sum_{i=1}^{n} | < v_i , \mu_j(B_i) > | \leqq 2^{-j}$$

for each finite partition $\{B_1 , \ldots , B_n\}$ of X with $B_i \in \mathcal{B}$, and each $\{v_1 , \ldots , v_n\} \subset V_j$. Define $W_j = V_1 + \ldots + V_j$. Then W_j is a convex, balanced 0-neighborhood in F_j and it follows from (**) that

$$(***) \qquad \sum_{i=1}^{n} | < w_i , \mu_j(B_i) > | \leqq \sum_{r=1}^{j} 2^{-r} \leqq 1$$

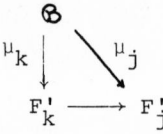

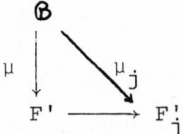

for each finite partition $\{B_1, \dots, B_n\}$ of X with $B_i \in \mathcal{B}$, and each $\{w_1, \dots, w_n\} \subset W_j$. Define $W = \cup W_j$. Since $W_j \subset W_{j+1}$ for every j, if follows that W is a convex, balanced 0 - neighborhood in F, and (*) holds, as asserted.

Since μ has total variation ≤ 1 with respect to q, it follows that the integral $\int \phi \, d\mu$ defines a continuous linear functional S on $\mathcal{C}(X;F_q)$ with $\|S\| \leq 1$, F_q being the vector space F, seminormed by q. But clearly S coincides with T_j on $\mathcal{C}(X;F_j)$ for every j. Hence S coincides with T on $\mathcal{C}(X;F)$. Thus

$$<\phi, T> = \int_X \phi \, d\mu$$

and

$$|<\phi, T>| \leq \sup_{x \in X} q(f(x))$$

for all $\phi \in \mathcal{C}(X;F)$. Thus T is τ_u - continuous, concluding the proof.

An examination of the proof of Lemma 1.3 shows that we can prove even more:

1.4 LEMMA. If $\mathcal{C}(X;F) = \cup \mathcal{C}(X;F_j)$, then a collection of linear forms on $\mathcal{C}(X;F)$ is τ_u - equicontinuous if and only if it is τ_i-equicontinuous.

Now we can prove

1.5 THEOREM. If $\mathcal{C}(X;F) = \cup \mathcal{C}(X;F_j)$, or in particular if the inductive limit $F = \text{ind lim } F_j$ is compactly - regular or Cauchy - regular, then the topologies τ_u and τ_i coincide.

PROOF. Theorem 1.5. follows from Lemmas 1.3 and 1.4, for any locally convex space has the topology of uniform convergence on the equicontinuous subsets of its dual space [12, Ch. IV, Th. 1.5, Cor. 3].

1.6 REMARK. If $\mathcal{C}(X;F) = \cup \mathcal{C}(X;F_j)$, then $\mathcal{C}(X;F)$ will be always endowed with the topology τ_u, or equivalently τ_i. We will then identify the continuous linear forms on $\mathcal{C}(X;F)$ with the F'-valued regular Borel measures μ on X which are of bounded variation with respect to some continuous seminorm on F, writting indifferently $<\phi, \mu>$ or $\int \phi d\mu$ for any $\phi \in \mathcal{C}(X;F)$.

We conclude this section with a characterization of the bounded subsets of $\mathcal{C}(X;F)$.

1.7 PROPOSITION. If the inductive limit $F = \text{ind lim } F_j$ is Cauchy-regular, then $\mathcal{C}(X;F) = \text{ind lim } \mathcal{C}(X;F_j)$ is a regular inductive limit.

PROOF. According to a result of Grothendieck [8, p. 78, Th. 9], each bounded subset of $\mathcal{C}(X;F)$ is contained in the closure of a bounded subset of some $\mathcal{C}(X;F_j)$. Thus to prove Proposition 1.7 we will show that the $\mathcal{C}(X;F)$-closure of the closed unit ball $\mathcal{B}_j$ of $\mathcal{C}(X;F_j)$ is contained in the closed unit ball $\mathcal{B}_k$ of $\mathcal{C}(X;F_k)$ for some $k \geqq j$.

If B_j denotes the closed unit ball of F_j, then, $F = \text{ind lim } F_j$ being Cauchy-regular, there exist r and k, with $j \leqq r \leqq k$, such that F and F_r induce the same Cauchy nets in B_j, and F and F_k induce the same Cauchy nets in B_r. Let ϕ belong to the $\mathcal{C}(X;F)$ - closure of $\mathcal{B}_j$. Then there exists a net (ϕ_α) in $\mathcal{B}_j$ which converges to ϕ in $\mathcal{C}(X;F)$. It follows that $\phi_\alpha(x)$ converges to $\phi(x)$ in F for each $x \in X$, and hence that $\phi_\alpha(x)$ converges to $\phi(x)$ in F_r for each $x \in X$. Since

$$\|\phi_\alpha(x)\|_r \leqq \|\phi_\alpha(x)\|_j \leqq \|\phi_\alpha\|_j \leqq 1$$

for every $x \in X$, if follows that $\|\phi(x)\|_r \leqq 1$ for every $x \in X$ and therefore $\phi(X) \subset B_r$. Then $\phi \in \mathcal{C}(X;F)$ implies that $\phi \in \mathcal{C}(X;F_k)$ and it follows that the $\mathcal{C}(X;F)$ - closure of $\mathcal{B}_j$ is contained in $\mathcal{B}_k$, concluding the proof.

2. SPACES OF SEQUENCES OF CONTINUOUS FUNCTIONS. Let X be a compact Hausdorff space and let (F_m) be a sequence of (LB)-spaces, $F_m = \text{ind lim } F_{mj}$. Let $\mathcal{S}_j$ be the Banach space of all sequences $\Phi = (\phi_m)$ such that, for every m, $\phi_m \in \mathcal{C}(X;F_{mj})$ and

$$\|\Phi\|_j = \sup_m j^{-m} \|\phi_m\|_{mj} < \infty .$$

Then $\mathcal{S}_j \hookrightarrow \mathcal{S}_{j+1}$ continuously for every j. We define $\mathcal{S} = \cup \mathcal{S}_j$ and endow $\mathcal{S}$ with the inductive topology with respect to the inclusion mappings $\mathcal{S}_j \hookrightarrow \mathcal{S}$.

We give a lemma on inductive limits whose straightforward proof we omit.

2.1 LEMMA. Let $G = \text{ind lim } G_\alpha$ be an inductive limit of subspaces with $G = \cup G_\alpha$. Let $\pi : G \to G$ be a projection such that $\pi(G_\alpha) \subset G_\alpha$ and $\pi | G_\alpha : G_\alpha \to G_\alpha$ is continuous for every α. Let $M = \pi(G)$ and let $M_\alpha = \pi(G_\alpha)$ for every α, endowed with the induced topologies of G and G_α, respectively. Then

(a) $M_\alpha = M \cap G_\alpha$, $M = \cup M_\alpha$.

(b) The identity mapping $M \to \text{ind lim } M_\alpha$ is a homeomorphism.

2.2 LEMMA. Assume $\mathcal{C}(X;F_m) = \cup\, \mathcal{C}(X;F_{mj})$ for every m and consider the linear mappings

$$\pi_m : \Phi = (\phi_m) \in \mathfrak{S} \to (0, \ldots, 0, \phi_m, 0, \ldots) \in \mathfrak{S}$$

$$\sigma_m : \phi \in \mathcal{C}(X;F_m) \to (0, \ldots, 0, \phi, 0, \ldots) \in \mathfrak{S}$$

Then:

(a) π_m is a continuous projection.

(b) $\Phi = \Sigma\, \pi_m(\Phi)$ for every $\Phi \in \mathfrak{S}$.

(c) σ_m is a homeomorphism between $\mathcal{C}(X;F_m)$ and $\pi_m(\mathfrak{S})$.

PROOF. (a) Clearly $\pi_m(\mathfrak{S}_j) \subset \mathfrak{S}_j$ and $\pi_m \mid \mathfrak{S}_j$ is a continuous projection on $\mathfrak{S}_j$.

(b) If $\Phi = (\phi_m) \in \mathfrak{S}_j$, then

$$\| \Phi - \sum_{m=0}^{N} \pi_m(\Phi) \|_{j+1} = \sup_{m \geqq N+1} (j+1)^{-m} \| \phi_m \|_{m,j+1}$$

$$\leqq \sup_{m \geqq N+1} (j/(j+1))^m\, j^{-m} \| \phi_m \|_{mj}$$

$$\leqq (j/(j+1))^{N+1} \| \Phi \|_j$$

Hence, as $N \to \infty$, $\sum_{m=0}^{N} \pi_m(\Phi)$ converges to Φ in $\mathfrak{S}_{j+1}$, hence in $\mathfrak{S}$.

(c) Certainly $\sigma_m \mid \mathcal{C}(X;F_{mj})$ is a homeomorphism between $\mathcal{C}(X;F_{mj})$ and $\pi_m(\mathfrak{S}_j)$, and it follows from Lemma 2.1 that σ_m is a homeomorphism between $\mathcal{C}(X;F_m)$ and $\pi_m(\mathfrak{S})$.

Next we characterize the dual of $\mathfrak{S}$ in terms of vector measures.

2.3 PROPOSITION. If $\mathcal{C}(X;F_m) = \cup\, \mathcal{C}(X;F_{mj})$ for every m , then given $T \in \mathfrak{S}'$ there exists a unique sequence of vector measures μ_m such that

(a) $\mu_m \in \mathcal{C}'(X;F_m)$;

(b) $\langle \Phi , T \rangle = \sum_m \int_X \phi_m d\mu_m$ for every $\Phi = (\phi_m) \in \mathfrak{S}$;

(c) if $\| \mu_m \|_{mj}$ denotes the norm of μ_m as a member of $\mathcal{C}'(X;F_{mj})$, then $\lim_{m\to\infty} \| \mu_m \|_{mj}^{1/m} = 0$ for every j .

Conversely, given a sequence of vector measures μ_m satisfying

(a) and (c), then (b) defines $T \in \mathfrak{S}'$.

PROOF. If σ_m is the mapping defined in Lemma 2.2, then we define $\mu_m \in \mathcal{E}'(X;F_m)$ by $\mu_m = T \cdot \sigma_m$. Then by Lemma 2.2 for any $\Phi = (\phi_m) \in \mathfrak{S}$ we get

$$<\phi, T> = \sum_m <\pi_m(\Phi), T> = \sum_m <\sigma_m(\phi_m), T> = \sum_m <\phi_m, \mu_m> = \sum_m \int_X \phi_m d\mu_m$$

Fix j. We will show that $\|\mu_m\|_{mj}^{1/m} \to 0$ as $m \to \infty$. There exists $\phi_m \in \mathcal{E}(X;F_{mj})$ such that

$$\|\phi_m\|_{mj} = 1, \quad <\phi_m, \mu_m> \geqq (1/2)\|\mu_m\|_{mj}$$

Let $\Phi_k = (k^m \phi_m)$. Then whenever $k \geqq j$, $\Phi_k \in \mathfrak{S}_k$ and

$$<\Phi_k, T> = \sum_m k^m <\phi_m, \mu_m> \geqq (1/2) \sum_m k^m \|\mu_m\|_{mj}$$

Thus $\sum_m k^m \|\mu_m\|_{mj} < \infty$ for every $k \geqq j$ and therefore $\|\mu_m\|_{mj}^{1/m} \to 0$ as $m \to \infty$.

The uniqueness of the sequence (μ_m) is clear.

To prove the converse, let (μ_m) be a sequence of vector measures satisfying (a) and (c). Then, given $\Phi = (\phi_m) \in \mathfrak{S}_j$, we have:

$$\sum_m |<\phi_m, \mu_m>| \leqq \sum_m \|\phi_m\|_{mj} \|\mu_m\|_{mj} \leqq \|\Phi\|_j \sum_m j^m \|\mu_m\|_{mj}$$

and the last written series converges for $\|\mu_m\|_{mj}^{1/m} \to 0$ as $m \to \infty$. Thus (b) defines a linear form T on $\mathfrak{S}$ whose restriction to each $\mathfrak{S}_j$ is continuous.

We conclude this section with a characterization of the bounded subsets of $\mathfrak{S}$.

2.4 PROPOSITION. Let the (LB) - spaces $F_m = \text{ind lim } F_{mj}$ have the following property: F_m and $F_{m,j+1}$ induce the same Cauchy nets in the closed unit ball of F_{mj} for every j and every m. Then the inductive limit $\mathfrak{S} = \text{ind lim } \mathfrak{S}_j$ is regular.

PROOF. The proof is similar to that of Proposition 1.7. Again by [8, p. 78, Th. 9] it suffices to show that the $\mathfrak{S}$ - closure of the closed unit ball $\mathcal{B}_j$ of $\mathfrak{S}_j$ is contained in the closed unit ball $\mathcal{B}_{j+2}$ of

$\mathcal{S}_{j+2}$. Let Φ belong to the $\mathcal{S}$-closure of $\mathcal{B}_j$ and let (Φ_α) be a net in $\mathcal{B}_j$ which converges to Φ in $\mathcal{S}$. Then $\phi_{\alpha m}(x)$ converges to $\phi_m(x)$ in F_m for every $x \in X$ and every m. Since $\|\Phi_\alpha\|_j \leqq 1$ it follows that $\|\phi_{\alpha m}(x)\|_{mj} \leqq j^m$ for every $x \in X$ and every m. Therefore, it follows from the hypothesis that $\phi_{\alpha m}(x)$ converges to $\phi_m(x)$ in $F_{m,j+1}$ for every $x \in X$ and every m. Thus $\phi_m(X) \subset F_{m,j+1}$ and $\|\phi_m(x)\|_{m,j+1} \leqq j^m$ for every $x \in X$ and every m. Then we show as in the proof of Proposition 1.7 that $\phi_m \in \mathcal{C}(X;F_{m,j+2})$ for every m. It follows that Φ belongs to $\mathcal{B}_{j+2}$, concluding the proof.

2.5 COROLLARY. Let the (LB) - spaces $F_m = \text{ind lim } F_{mj}$ have the following property: F_m and $F_{m,j+1}$ induce the same Cauchy nets in the closed unit ball of F_{mj} for every j and every m. Then, given $\mathcal{X} \subset \mathcal{S}$, the following conditions are equivalent:

(a) $\mathcal{X}$ is bounded in $\mathcal{S}$.

(b) $\mathcal{X}$ is contained and bounded in some $\mathcal{S}_j$.

(c) There exists j such that, for every $\Phi = (\phi_m) \in \mathcal{X}$, $\phi_m \in \mathcal{C}(X;F_{mj})$ and $\|\phi_m\|_{mj} \leqq j^m$, for every m.

3. THE SPACE OF HOLOMORPHIC GERMS. Throughout this section $\mathcal{H}(K)$ denotes the space of all complex-valued holomorphic germs on a compact subset K of a complex locally convex space E. E will be assume metrizable and Schwartz in our main result, namely Theorem 3.4. We refer to Grothendieck [8, p. 117, Def. 5] for the definition of Schwartz spaces. $\mathcal{H}(K)$ is endowed with the inductive topology coming from $\mathcal{H}(K) = \text{ind lim } \mathcal{H}^\infty(V)$, where $\mathcal{H}^\infty(V)$ denotes the Banach space of all complex-valued, bounded holomorphic functions on V, V varying among all open neighborhoods of K. $\mathcal{P}(^mE)$ denotes the space of all complex-valued continuous m-homogeneous polynomials on E. Our main references for $\mathcal{H}(K)$ and $\mathcal{P}(^mE)$ are [9], [5] and [1].

We begin with an intrinsecal characterization of the bounded subsets of $\mathcal{H}(K)$.

3.1 THEOREM. Let E be metrizable and let $K \subset E$ be compact and locally connected. Then, for any $\mathcal{X} \subset \mathcal{H}(K)$, the following conditions are equivalent:

(a) $\mathcal{X}$ is bounded in $\mathcal{H}(K)$.

(b) $\mathcal{X}$ is contained and bounded in $\mathcal{H}^\infty(V)$, for some open neighborhood V of K.

(c) There exists a continuous seminorm α on E and a constant $\rho > 0$ such that

$$\| \frac{1}{m!} \hat{d}^m f(x) \|_\alpha \leq \rho^{-m}$$

for every $f \in \mathfrak{X}$, every $x \in K$ and every m.

PROOF. The equivalence (a) $\Longleftrightarrow$ (b) is nothing but [9, Th. 3.1]. That (b) $\Longrightarrow$ (c) follows readily from the Cauchy inequalities. Thus we only have to prove that (c) $\Longrightarrow$ (b). The proof we will give is modelled on a proof of A. Baernstein [2, p. 31], who showed that (c) $\Rightarrow$ (b) in the case where $E = \mathbb{C}$.

For each $x \in K$ we choose $U_x \subset E$ open such that $x \in U_x \subset B_\alpha(x;\rho/2)$ and $K \cap U_x$ is connected, where $B_\alpha(x;r) = \{t \in E : \alpha(t-x) < r\}$.

For each $x \in K$ we choose next a continuous seminorm $\alpha_x \geq \alpha$ and V_x α_x - open with $x \in V_x \subset U_x$. Then we choose $x_1, \dots, x_n \in K$ such that $K \subset V_{x_1} \cup \dots \cup V_{x_n}$. Let $\beta = \max\limits_{1 \leq i \leq n} \alpha_{x_i}$. Then each V_{x_i} is β - open.

For each $x \in K$ we next choose V_{x_i} such that $x \in V_{x_i}$ and then choose $r_x > 0$ such that $B_\beta(x;2r_x) \subset V_{x_i}$. Thus

$$B_\beta(x\,;\,2r_x) \subset V_{x_i} \subset U_{x_i} \subset B_\alpha(x_i\,;\,\rho/2).$$

It follows that

$$B_\alpha(x_i\,;\,\rho/2) \subset B_\alpha(x\,;\,\rho).$$

Given $f \in \mathfrak{X}$ and $x \in K$ we define a function F_x on $B_\beta(x;r_x)$ by

$$F_x(t) = \sum_{m=0}^{\infty} \frac{1}{m!} \hat{d}^m f(x)(t-x)$$

Since $B_\beta(x\,;\,r_x) \subset B_\alpha(x\,;\,\rho/2)$ it follows that F_x is holomorphic and bounded by 2 on $B_\beta(x\,;\,r_x)$. Suppose for a moment that

$$(*) \qquad F_x(t) = F_y(t) \quad \text{for all} \quad t \in B_\beta(x\,;\,r_x) \cap B_\beta(y\,;\,r_y).$$

Then we can define a function F on

$$V = \bigcup_{x \in K} B_\beta(x;r_x)$$

by

$$F(t) = F_x(t) \quad \text{for} \quad t \in B_\beta(x\,;\,r_x).$$

Then F is holomorphic and bounded by 2 on V and f is the germ of F on K. Thus to conclude the proof if suffices the show (*).

We may assume $r_x \geq r_y$. If the intersection $B_\beta(x;r_x) \cap B_\beta(y;r_y)$ is nonvoid then $\beta(x-y) \leq r_x + r_y \leq 2r_x$ and hence $y \in B_\beta(x;2r_x) \subset U_{x_i}$. Thus x and y both belong to the connected set $K \cap U_{x_i}$. Let $U \supset K$ be an open set such that f is holomorphic on U. Then $K \cap U_{x_i} \subset U \cap B_\alpha(x;\rho)$. Let W be the connected component of $U \cap B_\alpha(x;\rho)$ which contains $K \cap U_{x_i}$. Then W is open and $x, y \in W$. By uniqueness of holomorphic continuation,

$$f(t) = \sum_{k=0}^{\infty} \frac{1}{k!} \hat{d}^k f(x)(t-x)$$

for every $t \in W$. Write

$$A_k = \frac{1}{k!} d^k f(x), \quad B_m = \frac{1}{m!} d^m f(y).$$

Then

$$f(t) = \sum_{k=0}^{\infty} \hat{A}_k(t-x)$$

for every $t \in W$ and consequently, by [10, §7, Prop. 3] we have that

$$\frac{1}{m!} d^m f(t) = \sum_{k=m}^{\infty} \frac{1}{m!} d^m \hat{A}_k(t-x)^{k-m}$$

for every $t \in W$, in particular

$$B_m = \frac{1}{m!} d^m f(y) = \sum_{k=m}^{\infty} \frac{1}{m!} d^m \hat{A}_k(y-x)^{k-m}$$

Consequently, for any $t \in B_\beta(x;r_x) \cap B_\beta(y;r_y)$, using [10, §5, Prop.1] we get

$$\begin{aligned} f_y(t) &= \sum_{m=0}^{\infty} B_m(t-y)^m \\ &= \sum_{m=0}^{\infty} \sum_{k=m}^{\infty} \frac{1}{m!} d^m \hat{A}_k(y-x)^{k-m}(t-y)^m \\ &= \sum_{k=0}^{\infty} \sum_{m=0}^{k} \binom{k}{m} A_k(y-x)^{k-m}(t-y)^m \\ &= \sum_{k=0}^{\infty} A_k(t-x)^k \\ &= f_x(t) \end{aligned}$$

concluding the proof.

3.2 REMARK. Theorem 3.1 was stated incorrectly by S. B. Chae for an arbitrary compact subset of a Banach spaces E [6,p.116,Prop.3.2] but R. Aron found a simple counterexample in $E = \mathbb{C}$.

3.3 REMARK. Theorem 3.1 has already been proved by R. L. Soraggi [15, Teor. 5.1] for a large class of compact sets which includes properly the locally connected ones, but his proof is much more involved.

We will characterize the dual of $\mathcal{H}(K)$ in terms of vector measures when K is a compact locally connected subset of a complex metrizable Schwartz space. In this situation, using [1, Lemma 4], we can find an increasing sequence of seminorms α_j defining the topology of E such that, letting E_j denote the vector space E seminormed by α_j, then $\mathcal{P}(^mE) = \text{ind lim}\, \mathcal{P}(^mE_j)$ and $\mathcal{P}(^mE_{j+1})$ induce the same Cauchy nets in the closed unit ball of $\mathcal{P}(^mE_j)$. Thus the (LB)-spaces $\mathcal{P}(^mE) = \text{ind lim}\, \mathcal{P}(^mE_j)$ are all Cauchy-regular, and even the stronger hypothesis in Proposition 2.4 or its Corollary 2.5 is fulfilled. With this notation we have:

3.4 THEOREM. Let K be a compact locally connected subset of a complex metrizable Schwartz space E. Then, given a continuous linear functional T on $\mathcal{H}(K)$, there exists a sequence of vector measures μ_m such that

(a) $\mu_m \in \mathcal{C}'(K; \mathcal{P}(^mE))$;

(b) $\langle f, T\rangle = \sum_m \int_K \frac{1}{m!} \hat{d}^m f \, d\mu_m$ for every $f \in \mathcal{H}(K)$;

(c) if $\|\mu_m\|_{mj}$ denotes the norm of μ_m as a member of $\mathcal{C}'(K; \mathcal{P}(^mE_j))$, then, for every j, $\lim_{m\to\infty} \|\mu_m\|_{mj}^{1/m} = 0$.

Conversely, given a sequence of vector measures μ_m satisfying (a) and (c), then (b) defines $T \in \mathcal{H}'(K)$.

PROOF. Let $\mathfrak{S} = \text{ind lim}\, \mathfrak{S}_j$ be the space of sequences of continuous functions defined by the compact set K and the (LB)-spaces $\mathcal{P}(^mE) = \text{ind lim}\, \mathcal{P}(^mE_j)$. We define an injective linear mapping $L : \mathcal{H}(K) \to \mathfrak{S}$ by $L(f) = \phi = (\phi_m)$, where

$$\phi_m = \frac{1}{m!} \hat{d}^m f \in \mathcal{C}(K; \mathcal{P}(^mE))$$

for every m. Using the Cauchy inequalities we see readily that L is well-defined and continuous. We claim that L is also open. To see this observe that:

(i) every closed bounded subset of $\mathcal{H}(K)$ is compact: this follows for $\mathcal{H}(K)$ is a Silva space, according to [5, Th. 7] or [1. Th. 4];

(ii) $\mathfrak{S}$ is a (DF)-space: this follows from [8, p. 78, Th. 9];

(iii) if $\mathfrak{X}$ is a bounded subset of $\mathfrak{S}$ then $L^{-1}(\mathfrak{X})$ is a bounded subset of $\mathcal{H}(K)$: this follows from Corollary 2.5 and Theorem 3.1.

Thus L is open according to Baernstein's open mapping theorem [2,p.29, Lemma], proving our claim.

Thus L is a topological isomorphism betweem $\mathcal{H}(K)$ and a subspace of $\mathfrak{S}$, and Theorem 3.4 follows from the Hahan - Banach theorem and Proposition 2.3.

4. INDUCTIVE LIMITS AND THE ε - PRODUCT. If G and H are locally convex spaces, then $G \varepsilon H$ denotes the ε - product of G and H [13], i.e. the space $\mathcal{L}_e(G'_c ; H)$ of all continuous linear mappings of G'_c into H, with the topology of uniform convergence on the equicontinuous subsets of G', where G'_c denotes the dual G' of G, endowed with the topology of uniform convergence on all compact subsets of G which are convex and balanced.

Let E be a Banach space and let $F = \text{ind lim } F_j$ be a compactly-regular Hausdorff (LB) - space. Then each inclusion mapping $\mathcal{L}_e(E'_c;F_j) \hookrightarrow \mathcal{L}_e(E'_c;F)$ is continuous, and using [13, p. 41, Prop. 8] one can prove that $\mathcal{L}(E'_c ; F) = \cup \mathcal{L}(E'_c ; F_j)$. Thus we may identify $E \varepsilon F$ with $\text{ind lim } E\varepsilon F_j$ algebraically and the identity mapping $\text{ind lim } E \varepsilon F_j \to E \varepsilon F$ is continuous. It is not known whether this mapping is a homeomorphism. Bierstedt and Meise have shown that this is the case whenever F is a nuclear space [4, p. 209, Bemerkung 13]. As an application of Theorem 1.5 we obtain another partial answer to this question.

4.1 PROPOSITION. If X is a compact Hausdorff space and $F = \text{ind lim } F_j$ is a compactly - regular Hausdorff (LB) - space, then the identity mapping $\text{ind lim } \mathcal{C}(X) \varepsilon F_j \to \mathcal{C}(X) \varepsilon F$ is a homeomorphism.

PROOF. By [11, Th. 8.6] the linear mapping

$$T \in \mathcal{L}(\mathcal{C}'_c(X);F_j) \to f \in \mathcal{C}(X ; F_j)$$

given by

$$f(x) = T(\delta_x) \qquad (x \in X),$$

where $\delta_x \in \mathcal{C}'(X)$ is the evaluation at x, is a topological isomorphism between $\mathcal{C}(X) \varepsilon F_j$ and $\mathcal{C}(X;F_j)$ for every j. Likewise for $\mathcal{C}(X) \varepsilon F$ and $\mathcal{C}_u(X;F)$. And since by Theorem 1.5 the identity mapping ind lim $\mathcal{C}(X;F_j) \to \mathcal{C}_u(X;F)$ is a homeomorphism, Proposition 4.1 follows.

4.2 REMARK. To conclude that $\mathcal{C}(X) \varepsilon F$ is topologically isomorphic to $\mathcal{C}_u(X;F)$ it is required in [11, Th. 8.6] that F be quasi-complete. But certainly the closed, convex, balanced hull of each compact subset of F is compact, and [11, Th. 8.6] still holds under this weaker hypothesis.

REFERENCES

1. P. AVILÉS - J. MUJICA, Holomorphic germs and homogeneous polynomials on quasi-normable metrizable spaces, Rend. Mat., to appear.

2. A. BAERNSTEIN II, Representation of holomorphic functions by boundary integrals, Trans. Amer. Math. Soc. 160 (1971), 27 - 37.

3. K.-D. BIERSTEDT - R. MEISE, Bemerkungen uber die Approximationseigenschaft lokalkonvexer Funktionenraume, Math.Ann. 209 (1974), 99 - 107.

4. K.-D. BIERSTEDT - R. MEISE, Induktive Limites gewichteter Raume stetiger und holomorpher Funktionen, J. reine angew.Math. 282 (1976), 186 - 220.

5. K.-D. BIERSTEDT - R. MEISE, Nuclearity and the Schwartz Property in the theory of holomorphic functions on metrizable locally convex spaces, Infinite dimensional holomorphy and applications, Notas de Matemática, North-Holland, Amsterdam, to appear.

6. S. B. CHAE, Holomorphic germs on Banach spaces, Ann. Inst. Fourier (Grenoble) 21 (1971), 107 - 141.

7. K. FLORET - J. WLOKA, Einfuhrung in die Theorie der lokalkonvexen Raume, Lecture Notes in Math. 56, Springer-Verlag, Berlin, 1968.

8. A. GROTHENDIECK, Sur les espaces (F) et (DF), Summa Brasiliensis, Math. 3 (1954), 67 - 123.

9. J. MUJICA, Spaces of germs of holomorphic functions, Doctoral Thesis, University of Rochester, 1974, to appear in Advances in Math.

10. L. NACHBIN, Topology on spaces of holomorphic mappings, Ergebnisse der Mathematik und ihrer Grenzgebiete 47, Springer-Verlag, Berlin, 1969.

11. J. B. PROLLA, Approximation of vector - valued functions, Notas de Matemática, North - Holland, Amsterdam, to appear.

12. H. H. SCHAEFER, Topological vector spaces, Graduate Texts in Mathematics 3, Springer-Verlag, Berlin, 1971.

13. L. SCHWARTZ, Théorie des distributions à valeurs vectorielles. I, Ann. Inst. Fourier (Grenoble) 7(1957), 1 - 142.

14. I. SINGER, Sur les applications linéaires intégrales des espaces de fonctions continues. I, Rev. Math. Pures Appl. 4(1959), 391 - 401.

15. R. L. SORAGGI, Partes limitadas nos espaços de germes de applica- ções holomorfas, Doctoral Thesis, Universidade Federal do Rio de Janeiro, 1976.

Instituto de Matemática
Universidade Estadual de Campinas
Caixa Postal 1170
13.100 - Campinas - SP - Brasil

LIFTINGS OF VECTOR MEASURES AND THEIR APPLICATIONS TO RNP AND WRNP

Kazimierz Musiał
Wrocław University and Polish Academy of Sciences

Czesław Ryll-Nardzewski
Technical University, Wrocław

Introduction.

Let B_1, B_2 be Banach spaces such that there exists a linear surjection $\varphi: B_1 \to B_2$.

The following factorisation problem can be considered: Does there exist for a given additive function $\nu: \mathcal{B} \to B_2$ ($\mathcal{B}$ is a Boolean algebra) an additive function $\varkappa: \mathcal{B} \to B_1$, such that $\varphi\varkappa = \nu$? Such a $\varkappa$ may be called a lifting of ν with respect to φ.

In this paper we consider the above problem in the case of $B_1 = X^*$ and $B_2 = Y^*$, for Banach spaces X and Y such that Y is a closed subspace of X and φ is the canonical surjection of X^* onto Y^*. We prove the existence of a lifting $\varkappa$ of ν provided ν is of finite variation. We show also that $\varkappa$ can be choosen to be of finite variation. Moreover, if ν is a measure (i.e. it is countably additive) then $\varkappa$ can also be choosen to be a measure.

We present two different proofs of the above statement. In one of them we use a Generalized Banach Limit which differs from the shift invariant Generalized Banach Limit (cf. Berberian [2]).

We consider also an analogous factorisation problem for Banach space valued operators defined on an L-space (section 3) and for Banach space valued measurable functions (section 5).

Using the existence of the lifting of vector measures of finite variation, we prove (section 4) that X does not contain any isomorphic copy of l_1, provided X^* has the weak Radon-Nikodym property (WRNP). This is a partial answer to a problem posed by Musiał in [7].

The first two sections of the paper are of introductory character and, may be, the results presented there are known.

Throught the paper the letters X,Y and Z denote Banach spaces (real or complex) and Y is always considered as a closed subspace of X with $T:Y \to X$ being the canonical isometric embedding.

(S,Σ,μ) always denotes a finite complete and positive measure space.

A function $f:S \to Z(Z^*)$ is <u>weakly</u> (<u>weak</u>*)-<u>measurable</u> with respect to μ provided $\langle f,u\rangle$ is measurable for every $u\in Z^*(u\in Z)$.

A weak* measurable $f:S \to Z^*$ is called <u>weak* uniformly bounded</u> (w^*.u.b.) if there exists a finite number M such that for every $z\in Z$, we have $|\langle z,f\rangle| \leq M\|z\|$ μ-a.e.

Functions $f,g:S \to Z(Z^*)$ are called <u>weakly</u> (<u>weak</u>*) <u>equivalent</u> with respect to μ if for every $u\in Z^*(u\in Z)$, we have $\langle u,f\rangle = \langle u,g\rangle$ μ-a.e.

The notions of a weak* scalarly integrable function, the Pettis integrability, weak* measure and that of the variation of an additive Banach space valued function defined on a Boolean algebra are standard If ν is an additive Z-valued function defined on a Boolean algebra $\mathcal{B}$ then the variation of ν on an element b from $\mathcal{B}$ is denoted by $|\nu|(b)$. ν is of finite variation if $|\nu|(1) < \infty$, where 1 is the unit element of $\mathcal{B}$.

1. <u>Generalized Banach Limits</u>.

Let us fix a nonempty set T directed upwards by a relation $\leq$.

For a completely regular topological (c.r.t.) space K we shall consider the collection $M(T,K)$ of all mappings $\{x_t\}_{t\in T}$ from T into K with conditionally compact ranges. The elements of $M(T,K)$ will be sometimes called K-valued nets.

PROPOSITION 1. <u>There exists a limit operation</u> Lim_t <u>defined on</u> $M(T,K)$ <u>simultaneously for all c.r.t. spaces</u> K <u>such that</u>:

1^0. <u>The value of</u> $\operatorname{Lim}_t x_t$ <u>is an accumulation point of the net</u> $\{x_t\}$;

2^0. <u>If</u> $x_t = y_t$ <u>for all sufficiently large</u> t, <u>then</u> $\operatorname{Lim}_t x_t = \operatorname{Lim}_t y_t$;

3^0. (<u>Consistency property</u>) <u>For any c.r.t. spaces</u> K <u>and</u> K', <u>we have</u> $\operatorname{Lim}_t \varphi(x_t) = \varphi(\operatorname{Lim}_t x_t)$, <u>provided</u> φ <u>is a continuous</u> K'-<u>valued mapping with its domain containing the closure of the set</u> $\{x_t : t\in T\}$.

Lim_t will be called a <u>Generalized Banach Limit</u>.

<u>REMARK</u> 1. It is easily seen that the operation Lim_t can also be defined for all nets of the form $\{x_t\}_{t\geq t_0}$, where t_0 is an arbitrary element of T. Then, we have $\operatorname{Lim}_t x_t = \operatorname{Lim}_{t\geq t_0} x_t$ and the properties

1^0-3^0 (suitably modified) are fulfilled.

REMARK 2. For every convergent net $\{x_t\}$ we have $\operatorname{Lim}_t x_t = \lim_t x_t$.

REMARK 3. If we take as a c.r. t. the space of reals R, then M(T,R) is the Banach space $l_\infty(T)$ and Lim_t is a linear, multiplicative, non-negative and normalized functional satisfying 1^0 and 2^0. The same holds for the complex numbers.

REMARK 4. For a given directed set T there is a natural one-to-one correspondence between Generalized Banach Limits and the set of the all accumulation points of the net $\{t\}_{t\in T}$ treated as an element of $M(T,\beta(T))$, where $\beta(T)$ is the Stone-Čech compactification of the discrete space T.

This remark indicates also a method of proving of Proposition 1.

2. Banach spaces of vector measures.

Assume that Ξ is an algebra of subsets of a set S.

We denote by $B(S,\Xi,Z)$ the space consisting of all uniform limits of Z-valued Ξ-measurable simple functions. Let $B(S,\Xi,Z)$ be endowed with the uniform norm: $|f| = \sup\{\|f(s)\| : s\in S\}$.

$ba(S,\Xi,Z)$ denotes the space of all additive set functions $\nu:\Xi \to Z$ of finite variation. We introduce in $ba(S,\Xi,Z)$ the variation norm: $|\nu| = |\nu|(S)$.

It is easily seen that $B(S,\Xi,Z)$ and $ba(S,\Xi,Z)$ are Banach spaces.

The following theorem is a natural generalization of Theorem IV.5.1. from Dunford and Schwartz [4]:

PROPOSITION 2. The formula

$$\langle x^*, f\rangle = \int_S f(s)\nu(ds)$$

(where the integral is understood in the sense of Bartle [1]) determines a linear isometry between the spaces $B^*(S,\Xi,Z)$ and $ba(S,\Xi,Z^*)$.

3. Liftings of vector measures.

The main result of this section is the following

THEOREM 1. *If $\mathcal{B}$ is a Boolean algebra and $\nu : \mathcal{B} \to Y^*$ is an additive function of finite variation, then there exists an additive function of finite variation $\varkappa : \mathcal{B} \to X^*$ such that:*

(i) $T^*\varkappa(b) = \nu(b)$ *for every* $b \in \mathcal{B}$;

(ii) $|\varkappa|(b) = |\nu|(b)$ *for every* $b \in \mathcal{B}$;

(iii) *If $\mathcal{B}$ is a Boolean σ-algebra and ν is a measure, then $\varkappa$ is also a measure.*

$\varkappa$ can be called a *norm-preserving lifting of ν with respect to* T^*.

REMARK 5. Since $\|T^*\| \leq 1$, we have always $|\nu|(b) \leq |\varkappa|(b)$, for every $b \in \mathcal{B}$, and consequently the equalities (ii) are equivalent to a single condition: $|\varkappa|(1) = |\nu|(1)$.

Proof. Since the variation of any measure with values in a Banach space is countably additive, the assertion (iii) is a trivial consequence of (ii).

We shall present two different methods of proving of (i) and (ii).

1st method. Let $\mathbb{B}$ be the family of all finite sub-algebras of $\mathcal{B}$ directed by the relation of inclusion.

For a given $\mathcal{B}_0 \in \mathbb{B}$ let $\{b_1, \dots, b_n\}$ be the set of all atoms of $\mathcal{B}_0$.

According to the Hahn-Banach theorem there exist elements $x_1^*, \dots, x_n^*$ in X^* such that

$T^* x_i^* = \nu(b_i)$ and $\|x_i^*\| = \|\nu(b_i)\|, i=1,\dots,n.$

In this way we are able do define an additive function

$$\varkappa_{\mathcal{B}_0} : \mathcal{B}_0 \to X^*$$

such that

$$T^*\varkappa_{\mathcal{B}_0} = \nu|\mathcal{B}_0 \quad (\text{the restriction of } \nu \text{ to } \mathcal{B}_0).$$

and

$$(1) \qquad |\varkappa_{\mathcal{B}_0}|(1) = |\nu|\mathcal{B}_0|(1) \leq |\nu|(1)$$

We shall apply now a Generalized Banach Limit to nets from $M(\mathcal{B},X^*)$ putting

$$\varkappa(b) = \operatorname{Lim}_{\mathcal{B}_0} \varkappa_{\mathcal{B}_0}(b) , \quad \text{for } b \in \mathcal{B} .$$

$\varkappa(b)$ is correctly defined since bounded subsets of X^* are weak* conditionally compact and the boundedness of the set $\{\varkappa_{\mathcal{B}_0}(b): \mathcal{B}_0 \in \mathbb{B}\}$ follows from condition (1).

It is not too hard to check that $\varkappa: \mathcal{B} \to X^*$ has the all required properties. In fact,(i) is a direct consequence of 3^0 , (ii) follows from inequality (1) , and the additivity of Lim for scalar valued nets (see Remark 3) yields the additivity of $\varkappa$.

2nd method. Let S be the Stone space of $\mathcal{B}$ and let $\hat{\mathcal{B}}$ be the algebra of clopen subsets of S. In an obvious way ν can be considered as a set function on $\hat{\mathcal{B}}$. Since S is 0-dimensional space, Z-valued $\hat{\mathcal{B}}$-measurable simple functions form a dense subset of the Banach space $C(S,Z)$ of Z-valued continuous functions defined on S. Hence, we have $C(S,Z) = B(S,\hat{\mathcal{B}},Z)$.

In view of Proposition 2 we may consider ν as an element of $C^*(S,Y)$.

Obviously $C(S,Y)$ is a closed subspace of $C(S,X)$ and so by the Hahn-Banach theorem there exists $\varkappa \in C^*(S,X)$ such that

$$(2) \qquad \|\varkappa\| = \|\nu\| \quad \text{and} \quad \varkappa | C(S,Y) = \nu$$

Using Proposition 2 again we can identify $\varkappa$ with an element of $ba(S,\hat{\mathcal{B}},X^*)$.

It is easily seen that (2) is equivalent to the assertion of Theorem 1.

As a corollary from the above theorem we get the existence of a norm-preserving lifting for linear operators on L-spaces:

THEOREM 2. _If_ L _is_ _an_ _L-space_ _and_ $U:L \to Y^*$ _is a bounded linear operator, then there exists an operator_ $V:L \to X^*$ _such that_ $T^*V = U$ _and_ $\|V\| = \|U\|$.

Proof. If $L = L_1(S,\Sigma,\mu)$ then U is uniquely determined by a vector measure $\nu:\Sigma \to Y^*$ of finite variation, given by the relation $\nu(E) = U(1_E)$ cf. Dinculeanu [3],p.   ).

Moreover,

$$(3) \qquad \|U\| = \sup\{ |\nu|(E)/\mu(E) \ : \ 0 < \mu(E) \}.$$

Using Theorem 1 we obtain a measure $\varkappa:\Sigma \to X^*$ being a norm-preserving lifting of ν with respect to T^*.

$\varkappa$ induces an operator $V:L \to X^*$ possessing all the required properties, in particular the equality $\|V\| = \|U\|$ follows from (ii) and (3) applied to V and $\varkappa$.

If L is an arbitrary L-space, then according to the representation theorem of Kakutani (see e.g.[9],p.465) L is isometrically isomorphic to an l_1-join of a collection of spaces $L_1(\mu_\alpha)$, where μ_α are finite measures.

We put $V = \bigoplus_\alpha V_\alpha$, where V_α is a norm-preserving lifting of $U_\alpha = U|L_1(\mu_\alpha)$.

Obviously, one has

$$\|V\| = \sup_\alpha \|V_\alpha\| = \sup_\alpha \|U_\alpha\| = \|U\|$$

4. <u>Applications of the lifting to the RNP and the WRNP</u>.

K.Musiał introduced in [7] a notion of a Banach space possessing the weak Radon-Nikodym property (WRNP) which was defined as follows: A Banach space Z has the WRNP provided for each (S,Σ,μ) and each μ-continuous measure $\nu:\Sigma \to Z$ of finite variation there exists a function $f:S \to Z$ such that

$$\nu(E) = \text{Pettis} - \int_E f d\mu , \qquad \text{for all } E\in\Sigma.$$

In this section, we give a few applications of Theorem 1 to the investigation of Banach spaces with the Radon-Nikodym or the weak Radon-Nikodym property.

THEOREM 3. <u>If</u> X^* <u>has the (weak) Radon-Nikodym property, then</u> Y^* <u>also does</u>.

<u>Proof</u>. Assume that X^* has the WRNP and take a μ-continuous measure $\nu:\Sigma \to Y^*$ of finite variation.

Then, in virtue of Theorem 1 there exists a norm-preserving lifting $\varkappa:\Sigma \to X^*$ of ν with respect to T^*.

By assumption, there is $f:S \to X^*$ such that

$$\varkappa(E) = \text{Pettis} - \int_E f d\mu \quad , \qquad E\in\Sigma.$$

This yields

$$\nu(E) = \text{Pettis} - \int_E T^* f d\mu \quad , \qquad E\in\Sigma,$$

and so Y^* has the WRNP.

If X^* has the RNP then f can be choosen to be strongly measurable, what yields the strong measurability of T^*f.

This completes the proof.

REMARK 6. The RNP-part of Theorem 3 has been proved, by a different method, by Stegall ([10], Corollary 6).

THEOREM 4. *If*
(a): X^* *has the weak Radon-Nikodym property,*
then
(b) X *does not contain any isomorphic copy of* l_1.

Proof. Assume that X^* has the WRNP and Y is a separable subspace of X.

In virtue of Theorem 3 Y^* has the WRNP and so, by a result of Musiał ([7], Theorem 3) Y does not contain any isomorphic copy of l_1.

This proves the assertion.

The mutual equivalence of the conditions (a) and (b) was formulated as Problem 5 in [7].

Ch. Stegall proved in [10] the following result:

THEOREM. *If* X^* *has the Radon-Nikodym property then each separable subspace of* X *has a separable dual.*

For separable X a proof, different from that of Stegall, was given by Musiał [6]. Using [6] and Theorem 3 we obtain a new proof of the above Theorem.

5. Liftings of vector functions.

In this section ρ is a lifting on $L_\infty(S,\Sigma,\mu)$ (see e.g.[5]). The lifting ρ uniquely determines a lifting (denoted also by ρ) on the space of all w^*.u.b. Z^*-valued functions (cf.[5], p.77).

If $f:S \to Y^*$ is a weak* measurable function then using the lattice properties of $L_1(S,\Sigma,\mu)$ we can find a sequence of pairwise disjoint sets $S_n \in \Sigma$ such that $S = \bigcup_{n\,0}^{\infty} S_n$, $\mu(S_0) = 0$ and for $n > 0$ $f|S_n$ is w^*.u.b. and $\rho(S_n) = S_n$.

In virtue of Proposition 7 of Dinculeanu ([3],p.215) there exists then a weak* measurable function $\tilde{f}:S \to Y^*$ such that for all $n>0$ and all $y \in Y$, we have,

$$\langle y,\tilde{f}\rangle|S_n = \rho(\langle y,f|S_n\rangle)$$

It is easily seen that $\tilde{f}$ is weak* equivalent to f and the norm of $\tilde{f}$ is a measurable function.

From now on we shall use the notation $\rho(h)$ for $\tilde{h}$ which is constructed for an arbitrary weak* measurable $h:S \to Y^*$(or X^*) , in the same way as $\tilde{f}$ was constructed from f, provided $h|S_n$ is w^*.u.b. on S_n, $n > 0$. Moreover, we shall assume always that $\rho(h)|S_0 = 0$.

PROPOSITION 3. If $f:S \to Y^*$ is weak* measurable then there exists a weak* measurable $g:S \to X^*$ such that $\|g\| = \|\rho(f)\|$ μ-a.e., $g = \rho(g)$, and $T^*g = \rho(f)$. If f is weak* scalarly integrable (weak* uniformly bounded) and the induced weak* measure $\nu:\Sigma \to Y^*$ given by

$$\langle y,\nu(E)\rangle = \int_E \langle y,f\rangle d\mu \quad , \text{ for all } y \in Y \text{ and } E \in \Sigma,$$

is of finite variation, then g can also be choosen to be weak* scalarly integrable (weak* uniformly bounded).

Proof. It is sufficient to consider the case of a weak* scalarly integrable f.

By assumption ν is of finite variation and so it is a measure in the norm topology of Y^*.

According to Theorem 1 there exists a norm-preserving lifting $\varkappa:\Sigma \to X^*$ of ν with respect to T^*.

There exists a weak* scalarly integrable $g:S \to X^*$ such that $g = \rho(g)$ and

$$\langle x,\varkappa(E)\rangle = \int_E \langle x,g\rangle d\mu$$

for every $E \in \Sigma$ and $x \in X$. The existence of g follows easily from a representation theorem of A.and C. Ionescu Tulcea (see e.g.[5],Chapter VII, Theorem 1) and the explicite formulation of that statement can be found in Dinculeanu ([3],§13) and Rybakov [8].

Since (cf. Musiał [7], Proposition 1)

$$|\nu|(E) = \int_E \|\rho(f)\| d\mu$$

and

$$|\varkappa|(E) = \int_E \|g\| d\mu ,$$

we get the equality $\|\rho(f)\| = \|g\|$ μ-a.e.

It is easily seen that T^*g and f are weak* equivalent and therefore $\rho(T^*g) = \rho(f)$. Moreover, it is not difficult to check that $\rho(T^*g) = T^*g$ and consequently $T^*g = \rho(f)$, what completes the proof.

g is called a lifting of f with respect to T^*.

For weakly measurable functions the following theorem holds:

PROPOSITION 4. *If X^* has the weak Radon-Nikodym property then for any Pettis integrable function $f:S \to Y^*$ inducing a measure $\nu:\Sigma \to Y^*$ of finite variation by the formula*

$$\nu(E) = \text{Pettis} - \int_E f d\mu \quad , \quad E\in\Sigma,$$

there exists a Pettis integrable function $g:S \to X^*$ such that T^*g and f are weakly equivalent (if Y is separable then they are simply equal μ-a.e.). If X^* has the Radon-Nikodym property then g can be taken to be strongly measurable.

Proof. Assume that X^* has the WRNP and $\nu:\Sigma \to Y^*$ is the measure being the Pettis integral of f.

In virtue of Theorem 1 there is a norm- reserving lifting $\varkappa:\Sigma \to X^*$.

Since X^* has the WRNP, there exists a function $g:S \to X^*$ such that

$$\varkappa(E) = \text{Pettis} - \int_E g d\mu \quad , \qquad E\in\Sigma.$$

Because of the equalities

$$\int_E f d\mu = \nu(E) = T^*\varkappa(E) = \int_E T^* g d\mu$$

we get the weak equivalence of f and T^*g.

We are grateful to Professor N. Kalton for his information that theorem 2 can be obtained also as a corollary from some earlier results c.f. Lacey "The isometric theory of classical Banach spaces" p.178.

Generalized Banach limits for ordinary sequences have been considered by F. Mazur and others c.f. [9] p.246.

REFERENCES

[1] Bartle R.G., A general bilinear vector integral, Studia Math. 15(1956),337-352.

[2] Berberian S.K., Lectures in Functional Analysis and Operator Theory, Graduate Texts in Math.,vol.15, Springer-Verlag (1974).

[3] Dinculeanu N., Vector measures, International Ser.of Monographs in Pure and Appl.Math.,vol.95, Pergamon Press (1967).

[4] Dunford N.S. and Schwartz J.T., Linear operators I.,Interscience,New York (1958).

[5] Ionescu Tulcea A. and Ionescu Tulcea C., Topics in the Theory of Lifting, Ergebnisse der Math.und ihrer Grenzgebiete, vol 48, Springer-Verlag (1969).

[6] Musiał K., Functions with values in a Banach space possessing the Radon-Nikodym property, Aarhus Universitet (1977) Preprint Series No.29,

[7] Musiał K., The weak Radon-Nikodym property in Banach spaces, Studia Math. to appear .

[8] Rybakov V.I., On vector measures in Russian , Izviestija Vysših Učebnyh Zaviedenij, Matiematika 79(1968),92-101.

[9] Semadeni Z., Banach spaces of continuous functions,vol.I. PWN-Polish Scientific Publishers, Warszawa (1971).

[10] Stegall Ch., The Radon-Nikodym property in conjugate Banach spaces, Trans. Amer. Math. Soc. 206(1975), 213-223.

50-384 Wrocław
Pl.Grunwaldzki 2/4
Poland

50-370 Wrocław
W.Wyspiańskiego 27
Poland

INTEGRAL REPRESENTATIONS IN CONUCLEAR SPACES

by

Erik Thomas

The University of Groningen

Netherlands

Summary. The main result can be formulated as follows: Let Γ be a closed convex cone in a quasi-complete conuclear space whose dual contains a countable total subset (e.g. $\mathbb{R}^{\mathbb{N}}$, $\mathcal{D}'$, $\mathcal{E}'$, $\mathcal{S}'$, any nuclear Frechet space). Assume that Γ satisfies condition $(\mathcal{C})$: For every compact convex set $K \subset \Gamma$, the set of points 'between' the origin and points of K (i.e. $\Gamma \cap K - \Gamma$) is compact. Then every point of Γ has an integral representation by means of extreme generators of Γ; this integral representation is unique if and only if Γ is a lattice.

More precisely: let $\mathcal{S}$ be a subset of Γ, not containing the origin, having one point on each ray, and such that $\bigcup_{0 \leq \lambda \leq 1} \lambda \mathcal{S}$ is a Borel set (such sets can be constructed). Let $\mathcal{S}_e$ be the set of points of $\mathcal{S}$ lying on extreme rays of Γ. Let $M^+(\mathcal{S}_e)$ be the set of Radon measures on $\mathcal{S}_e$, having finite first moments, such that the weak integrals $\int x dm(x)$ belong to Γ. Then A) for every point $a \in \Gamma$ there exists $m \in M^+(S_e)$ such that $a = \int x dm(x)$. B) for every $a \in \Gamma$ there is precisely one $m \in M^+(\mathcal{S}_e)$ such that $a = \int x dm(x)$, if and only if Γ is a lattice.

The condition $(\mathcal{C})$ implies that Γ is proper $(\Gamma \cap -\Gamma = (o))$. Any weakly complete proper convex cone in a quasi-complete conuclear space satisfies condition $(\mathcal{C})$. Other examples are given.

In the last part of the paper we propose an abstract generalization: 'conuclear' cones. The cones in a conuclear space satisfying condition $(\mathcal{C})$, and the cones with compact base in an arbitrary space are both conuclear.

Method. In order to prove these results we work, not with Radon measures on the rather arbitrary sets $\mathcal{S}$ defined above, but with the notion of conical measure defined by G. Choquet. We restrict ourselves, however, to those conical measures which can be defined by means of integrals: localizable conical measures.

Contents

1. Localizable conical measures

Let F be a locally convex Hausdorff space over $\mathbb{R}$. Let S be the set of finite suprema $\varphi = \sup_i \ell_i$, with $\ell_i \in F'$ and let h(F) be the linear function space S - S composed of all differences of such functions.

Recall that a conical measure on F is a linear form μ on h(F) such that $\mu(\varphi) \geq 0$ if $\varphi(x) \geq 0$ for all $x \in F$ (cf. G. Choquet [1], [2], [3]).

We denote by Γ a closed convex cone in F. If $\mu(\varphi) \geq 0$ for every $\varphi \in h(F)$ such that $\varphi \geq 0$ on Γ, μ is said to be concentrated on Γ. From now on we shall deal only with conical measures (concentrated) on Γ. We shall assume, moreover, that Γ is countably separated, i.e. that F' contains a countable subset separating the points of Γ.

Definition 1. A conical measure μ on Γ is localizable if there exists a Radon measure m on $\Gamma_* = \Gamma \setminus \{0\}$, with finite first moments, such that $\mu(\varphi) = \int \varphi dm$ for all $\varphi \in h(F)$. The measure m will be called a localization of μ.(1)

If μ is localizable there exist infinitely many localizations: e.g. $\mu(\varphi) = \lambda^{-1}\int \varphi(\lambda x)\,dm(x)$ for all $\lambda > 0$, the functions $\varphi \in h(F)$ being positively homogeneous of degree one.

If however S is a subset of Γ_* which encounters each ray, issuing from the origin, in at most one point (we shall call such a set a sky) it can be shown that μ has at most one localization concentrated on S.(2)

Conversely, it is possible to construct a sky $S \subset \Gamma$ with the property that every localizable conical measure concentrated on Γ has a, necessarily unique, localization on S. (If there exists $\ell \in F'$ such that $\ell(x) > 0$ for all $x \in \Gamma_*$ one can take $S = \{x \in \Gamma : \ell(x) = 1\}$).

Definition 2. Let μ be a localizable conical measure on Γ. A function $f : \Gamma \to \mathbb{R}$ is μ-integrable if 1. f is positively homogeneous of degree one, and 2. f is m-integrable, m being any localization of μ. In this case f is integrable with respect to any other localization and the integral $\int f dm$ is independent of the localication m; we put $\mu(f) = \int f d\mu = \int f dm$.

Definition 3. Let Γ_o be any cone in F. Then μ is said to be concentrated on Γ_o if some localization of μ is concentrated on Γ_o, in which case every other localization has the same property.

If μ_o is a closed convex cone the present definition coincides with the previous one.

---

(1) By Radon measure we mean a non negative locally finite measure on the Borel sets of a topological space, such that $m(A) = \sup\{m(K) : K \subset A, K \text{ compact}\}$. (cf. L. Schwartz [6]). Such general localizations have also been considered by Professor Choquet in his lectures at the University of Maryland in 1971.

(2) m is concentrated on a set $A \subset \Gamma$ if there exists a Borel set B containing $\Gamma \setminus A$ such that $m(B) = 0$. (The assumption on Γ implies that Γ is the countable union of open sets of finite m measure).

The assertions embodied in these definitions can be verified, roughly speaking, by 'sliding' two given localizations along rays to an appropriate sky, where, by the uniqueness property, the resulting localizations must coincide.

The resultant $r(\mu)$ of a conical measure μ is the point a of the weak completion of F such that $\ell(a) = \mu(\ell)$ for all $\ell \in F'$. Therefore, if m is a localization of μ we have $a = \int x\,dm(x)$, in the sense of weak integrals, which we may also write as $a = \int x\,d\mu(x)$.

Thus localizable conical measures are mathematical objects which can describe, in an intrinsic manner, integral representations in closed convex cones. In fact, it would be reasonable to say that an integral representation of a point $a \in \Gamma$, by means of extreme generators <u>is</u> a localizable conical measure, concentrated on the cone $\mathrm{ext}(\Gamma)$ of extreme rays of Γ, such that $a = \int x\,d\mu(x)$.

We shall denote by $M^+(\mathrm{ext}\Gamma)$ the set of localizable conical measures μ, concentrated on $\mathrm{ext}(\Gamma)$, such that $r(\mu)$ belongs to F (hence to Γ).

2. Integral representation in conuclear spaces

We assume now that Γ is a closed convex proper cone in F (i.e. $\Gamma \cap -\Gamma = (0)$). For any convex compact set $K \subset \Gamma$ we denote by K^* the set $\bigcup_{x \in K} [o,x]$, where $[o,x] = \Gamma \cap x - \Gamma$ is the <u>order interval</u> between o and x. It is easy to see that K^* is again closed and convex. But K^* need not be compact (not even the order interval $[o,x]$ need be compact).

<u>Theorem 1</u>. Let F be a quasi-complete conuclear space. Let Γ be a countably separated closed convex proper cone in F satisfying the following condition:

(C) For every convex compact set $K \subset \Gamma$, K^* is compact.

Then:

A) Even point $a \in \Gamma$ is the resultant of at least one conical measure $\mu \in M^+(\mathrm{ext}\Gamma)$.

B) Every $a \in \Gamma$ is the resultant of precisely one $\mu \in M^+(\mathrm{ext}\Gamma)$ if and only if Γ is a lattice.

<u>Corollary</u> Γ is the closed convex hull of $\mathrm{ext}(\Gamma)$.

<u>Remarks</u> 1. Under the hypothesis of theorem 1 every closed convex subcone of Γ also satisfies the condition (C), and is countably separated.

2. The space F being quasi-complete and conuclear every closed bounded set is compact. Thus, to verify condition (C) it is sufficient to check that K^* is bounded. This implies:

3. If there exists a family of continuous semi-norms which are additive on Γ, and such that the associated topology is compatible with F', condition (C) is satisfied. [3]

---

(3) e.g. in the cone of totally or absolutely monotone functions on an interval (a,b) in $\mathbb{R}$ the seminorms $u \to \int_K |u^{(n)}|\,dx$ have this property; K is a compact interval in (a,b).

In particular:

4. If every $\ell \in F'$ is the difference of two positive linear forms condition (C) is satisfied. More particularly, if Γ is weakly complete condition (C) is satisfied (cf. Choquet [3] 30-10).

5. Let F be contained, with continuous inclusion, in a locally convex Hausdorff space $\tilde{F}$. Assume Γ is a hereditary subcone of a weakly complete proper convex cone $\tilde{\Gamma} \subset \tilde{F}$. Then, if Γ is a Suslin subspace of F, condition (C) is satisfied (and Γ is countably separated).

In effect, Γ being hereditary in $\tilde{\Gamma}$, K^* is the same, calculated in Γ or $\tilde{\Gamma}$. Then by the previous remark K^* is weakly compact in $\tilde{F}$, and so is the set $A = \mathrm{co}\,(K^* \cup -K^*)$. Thus F_A is a Banach space which is continuously included in $\tilde{F}$, and it is a subspace of the locally convex Suslin space $F_o = \Gamma - \Gamma \subset \tilde{F}$. Consequently, by the closed graph theorem (L. Schwartz [6], p. 160) the inclusion of F_A in F_o is continuous, which means that A, hence K^*, is bounded in F, and consequently compact.

Example 1. Let V be a differentiable manifold of class C^∞. Let $F = \mathcal{D}'(V \times V)$. Then F is a quasi-complete conuclear Suslin space, in particular countably separated. Let Γ be the cone of distributions of positive type i.e. for which $\langle T, \alpha \otimes \bar{\alpha} \rangle \geq 0$ for all $\alpha \in \mathcal{D}$. Then the theorem applies to Γ and its closed subcones, (e.g. if V is a Lie group, to the set of kernels invariant under left and right translation (traces)).

Indeed, by the Schwartz kernel theorem, Γ can be identified with the cone of reproducing kernels of Hilbert subspaces of $\mathcal{D}'(V)$. Let $\tilde{\Gamma}$ be similarly the cone of reproducing kernels of Hilbert subspaces of the weak completion $\mathcal{D}^*(V)$. Then, any Hilbert subspace of a Hilbert subspace of $\mathcal{D}'$ being itself a Hilbert subspace of $\mathcal{D}'$, Γ is hereditary in $\tilde{\Gamma}$ and the conclusion follows from the above criterium (cf. L. Schwartz [7]).

2. The cone $\mathcal{D}_+$ of all non negative Schwartz test functions on $\mathbb{R}$ is an example of a cone in a quasi complete conuclear space, viz. $\mathcal{D}'$, which does not satisfy condition (C); indeed the order intervals are not even bounded. $\mathcal{D}_+$ does not have any extreme rays.

3. Proof of theorem 1

The definition of conuclearity most appropriate here is the following: For every closed bounded symmetric set $A \subset F$, there exists another $B \supset A$, such that the inclusion of the space F_A into F_B is absolutely summing (cf. A. Pietsch[5]). Equivalently: if $P_B(x) = \inf \{\lambda \geq 0 : x \in \lambda B\}$, there exists a constant M such that, for any finite family $(x_i)_{i \in I}$, for which $\sum_{i \in J} x_i \in A$ for all $J \subset I$, we have $\sum_{i \in I} P_B(x_i) \leq M$. (Replacing B by MB we can always assume M = 1).

Let Σ be the set of all convex compact subsets $A \subset \Gamma$ such that $A = A^*$.

Lemma 1. For any $A \in \Sigma$ there exists $B \in \Sigma$, such that $x_i \in \Gamma$ and $\sum_{i \in I} x_i \in A$ implies

$\sum_{i \in I} P_B(x_i) \leq 1.$

This follows easily from the conuclearity and condition (*C*). (The definition of P_B in this case is the same as before).

Recall the Choquet ordering between conical measures: $\nu < \mu$ if $\nu(\varphi) \leq \mu(\varphi)$ for all $\varphi \in S$.

We denote by D the set of conical measures of the form $\nu = \sum_{i \in I} \varepsilon_{x_i}$, with $x_i \in \Gamma$ and I finite, and by D_a those $\nu \in D$ for which $r(\nu) = a$, i.e. $\sum_{i \in I} x_i = a$.

Lemma 2. Let μ be a conical measure concentrated on Γ. Then the following three conditions are equivalent:

1. For every $\varphi \in S$, $\mu(\varphi) = \sup \{\nu(\varphi) : \nu < \mu, \nu \in D\}$.
2. μ is localizable on a set $B \in \Sigma$.(4)
3. μ is localizable and $r(\mu) \in \Gamma$.

Proof 1. implies 2: For $\nu \in D$ with $\nu < \mu$, $r(\mu) = r(\nu) \in \Gamma$.
Let $r(\mu)$ belong to $A \in \Sigma$ and let B be a set associated with A as in lemma 1. Define $\mu(P_B) = \sup \{\mu(\varphi) : \varphi \leq P_B, \varphi \in S\}$. Then, observing that, by the Hahn Banach theorem, $P_B = \sup\{\varphi \in S : \varphi \leq P_B\}$, we obtain the relations:

$$\mu(P_B) = \sup_{\substack{\nu < \mu \\ \nu \in D}} \nu(P_B) = \sup_{\sum \varepsilon_{x_i} < \mu} \sum_i P_B(x_i) \leq 1.$$

Let μ_1 be a conical measure such that $0 \leq \mu_1 \leq \mu$. Then we have, a fortiori, $\mu_1(P_B) \leq 1$. Let $\ell(x) \leq 1$ for all $x \in B$. Then $\ell \leq P_B$ and $\ell \in S$, consequently $\ell(r(\mu_1)) = \mu_1(\ell) \leq \mu_1(P_B) \leq 1$, whence $r(\mu_1) \in B^{oo} = B$. Thus the set $K_\mu = \{r(\mu_1) : 0 \leq \mu_1 \leq \mu\}$ is contained in Γ. This implies that μ is the limit of a net $\nu_\alpha \in D$, with $\nu_\alpha < \mu$ (cf. Choquet [3], 30.9). Now if $\nu_\alpha = \sum_i \varepsilon_{x_i} < \mu$, the inequality $\sum_i P_B(x_i) \leq 1$ shows that ν_α is localizable on B in the measure $\sum_i P_B(x_i)\delta_{x_i/P_B(x_i)}$, of total mass ≤ 1. Then by using the compactness, in the vague topology, of the set of Radon measures on B with total mass at most 1, one readily obtains a localization m of μ on B for which $\int_B dm \leq 1$.

2. implies 3. obvious.

3. implies 1. This follows by approximating the integrals by Riemann sums.

We denote by $M^+ = M^+(\Gamma)$ the set of conical measures concentrated on Γ,

---

(4) Here we follow the definition of G. Choquet viz. there exists a Radon measure m on B, in particular with finite total mass, such that $\mu(\varphi) = \int_B \varphi \, dm$. In this case μ is a fortiori localizable in the sense of definition 1.

possessing the properties described in lemma 2, and by M_x^+ the set of those $\mu \in M^+(\Gamma)$ for which $r(\mu) = x$.

For any $\varphi \in h(F)$ and $x \in \Gamma$ we define:

$$\varphi'(x) = \sup_{\mu \in M_x^+} \mu(\varphi)$$

Lemma 3 i) $\varphi(x) \leq \varphi'(x) < +\infty \quad \forall\, x \in \Gamma$.

ii) $\varphi'(x + y) \geq \varphi'(x) + \varphi'(y) \quad \forall\, x,y \in \Gamma$

iii) $\varphi'(\lambda x) = \lambda\varphi'(x) \quad \forall\, \lambda > 0 \quad \forall\, x \in \Gamma$.

iv) $(\varphi + \psi)'(x) \leq \varphi'(x) + \psi'(x) \quad \forall\, \varphi, \psi \in h(F)$.

v) $\varphi(x) = \varphi'(x) \quad \forall\, \varphi \in -S \quad \forall\, x \in \Gamma$.

vi) $\varphi(x) = \varphi'(x)$ for all $\varphi \in S$ if and only if x lies on an extreme ray.

vii) For every $A \in \Sigma$ the restriction $\varphi'|_A$ is upper semi-continuous. In particular φ' is μ-integrable for all $\mu \in M^+(\Gamma)$.

Proof. The first inequality in i) and the relations ii) to v) are obvious. The last inequality in i) and assertion vii) follow from the fact, established in connection with lemma 2, that for any $x \in A$ and $\mu \in M_x^+$, μ can be localized on a set B, independent of x, in a measure with total mass ≤ 1. Finally, vi) follows easily from the fact that for $\varphi \in S$, $\varphi'(x) = \sup_{\nu \in D_x} \nu(\varphi)$.

Lemma 4. $M^+(\Gamma)$ is inductive in the Choquet ordering.

This follows from lemma 3 i) and the characterization 1 in lemma 2.

Lemma 5. Let $\mu \in M^+(\Gamma)$; Let ν be any conical measure such that:
1. $\nu(\varphi) \leq \mu(\varphi')$ for all $\varphi \in h(F)$. Then we have 2. $\mu < \nu$ and 3. $\nu \in M^+(\Gamma)$.
Conversely, if Γ is a lattice, 2. and 3. imply 1.

Proof 1. implies 2. by lemma 3 v). Let $r(\mu) \in A \in \Sigma$; let B be associated with A, and C with B as in lemma 1. Then it is not hard to see that $\nu(P_C) \leq 1$; consequently ν is localizable on C (cf. the proof of lemma 2). For the converse we resort to the fact that, if Γ is a lattice, μ and ν can be approximated by nets μ_α and ν_α in D, such that $\mu_\alpha < \nu_\alpha$.

Proof of A). Let μ be maximal in $M^+(\Gamma)$, majorizing ε_a (lemma 4). Then, the functional $\varphi \to \mu(\varphi')$ being positively homogeneous and subadditive (Lemma 3, iii), iv) vii)), there exists, for every $\varphi_o \in h(F)$, a linear from ν on $h(F)$ such that $\nu(\varphi) \leq \mu(\varphi')$ for all $\varphi \in h(F)$, with equality for $\varphi = \varphi_o$ (Hahn Banach theorem). In particular, for $\varphi \leq 0$ $\nu(\varphi) \leq 0$; therefore ν is a conical measure. By lemma 5, $\mu < \nu$ and $\nu \in M^+(\Gamma)$, Hence, by the maximality of μ, $\nu = \mu$, whence $\mu(\varphi_o) = \nu(\varphi_o) = \mu(\varphi_o')$. This implies that μ is concentrated on the cone $B_{\varphi_o} = \{x \in \Gamma : \varphi_o(x) = \varphi_o'(x)\}$.
Now $\mathrm{ext}(\Gamma) = \bigcap_{\varphi \in S} B_\varphi$ (lemma 3 vi)). Similarly, if Γ is countably separated there

exists a countable system $(\ell_n)_n$ of continuous linear forms such that $ext(\Gamma) = \bigcap_n B_{\varphi_n}$, where $\varphi_n = |\ell_n|$. Thus μ is concentrated on $ext(\Gamma)$, (which proves that $ext(\Gamma) \neq (o)$).

Proof of B). It is not hard to see that the one $M^+(ext\Gamma)$ is hereditary in the cone of all conical measures by making use of the characterization 2 in lemma 2. Consequently $M^+(ext\Gamma)$ is a lattice. The uniqueness of the representation implies that Γ is linearly isomorphic to this lattice.

Conversely, assume Γ is a lattice. Then, for $\varphi \in S$, $\varphi'(a) = \lim_{\nu \in D_a} \nu(\varphi)$, the set D_a being now directed with respect to the Choquet ordering, by virtue of the decomposition property of Γ. Thus we may define a conical measure μ_a by putting, for any $\varphi \in h(F)$, $\mu_a(\varphi) = \lim_{\nu \in D_a} \nu(\varphi)$. Then μ_a is concentrated on Γ, and satisfies condition 1 in lemma 2 by construction; thus μ_a belongs to $M^+(\Gamma)$, and moreover $r(\mu_a) = a$. For $\varphi \in S$, $\mu_a(\varphi) = \varphi'(a)$; thus, by definition of $\varphi'(a)$, we have $\mu < \mu_a$ for all $\mu \in M^+_a$. Now assume that $\mu \in M^+_a$ is concentrated on $ext(\Gamma) = \bigcap_{\varphi \in S} B_\varphi$. Then we obtain, by the converse in lemma 5, the relations $\mu_a(\varphi) \leq \mu(\varphi') = \mu(\varphi)$ for all $\varphi \in S$; consequently $\mu_a < \mu$ whence $\mu = \mu_a$. The proof is complete.

Remarks 1. It is possible, under the hypotheses of theorem 1, to give a constructive existence proof by a method analogous to that of Hervé [4].

2. Under the hypotheses of theorem 1 $ext(\Gamma)$ is universally measurable. This results from the proof.

3. If Γ is not countably separated the present proof yields an analogous theorem where $M^+(ext\Gamma)$ is replaced by $\bigcap_\varphi M^+(B_\varphi)$.

4. The conical measures which are maximal in $M^+(\Gamma)$ need not be maximal in the cone $\mathcal{M}^+(\Gamma)$ of all conical measures carried by Γ. If, however, Γ is weakly complete, we have $M^+(\Gamma) = \mathcal{M}^+(\Gamma)$. (Every $\mu \in \mathcal{M}^+(\Gamma)$ then has the property that $K_\mu \subset \Gamma$, and so satisfies condition 1 in lemma 2).

This leads to the following theorem, the first part of which has been proved by G. Choquet:

Theorem 2. Let F be a locally convex Hausdorff space and let $\Gamma \subset F$ be a weakly complete proper convex cone.

A) Every point $a \in \Gamma$ is the resultant of a maximal conical measure $\mu \in \mathcal{M}^+(\Gamma)$, unique for each a if and only if Γ is a lattice.

B) If F is a conuclear space these conical measures (and all conical measures carried by Γ) are localizable on compact subsets of Γ.

C) If, moreover, Γ is countably separated a conical measure $\mu \in \mathcal{M}^+(\Gamma)$ is maximal if and only if it is concentrated on $ext(\Gamma)$.

B) and C) are consequences of what precedes except for the 'if' in C) which can easily be proved by the usual techniques because in this case the function φ' is upper semi-continuous throughout the entire cone.

4. A generalization

Let F be any quasi-complete locally convex Hausdorff space and let Γ be a closed convex proper cone in F.

Let Σ be a set of convex compact subsets of Γ.

Definition. We shall say that Γ is Σ-conuclear if the following conditions are satisfied:

1. $A = A^*$ for every $A \in \Sigma$, 2. $\Gamma = \bigcup_{A \in \Sigma} A$ and 3. the condition in lemma 1 is satisfied.

Examples 1. Every convex cone with a compact base is conuclear. More generally:

2. Every closed convex well capped cone is K-conuclear, K being the set of all caps of Γ.

Indeed, caps have the property that $A = A^*$, and the condition in lemma 1 is satisfied with $B = A$, the gauge P_A of a cap being additive. (Recall that a cap is a convex compact subset $A \subset \Gamma$ containing o such that $\Gamma \setminus A$ is convex).

3. Any closed convex cone Γ in a quasi-complete conuclear space, satisfying condition (C) is Σ-conuclear, Σ being the set of all convex compact subsets of Γ such that $A = A^*$.

Theorem 3. Let Γ be an Σ-conuclear cone in a quasi complete space F. Assume Γ to be countably separated. Then the conclusions A) and B) in theorem 1 are valid.

The proof is the same as for theorem 1.

Remark. It is not known whether every Σ-conuclear cone is actually well capped.

References

[1] G. Choquet, Les cônes convexes faiblement complets dans l'Analyse. Proc.Intern. Congress Mathematicians. Stockholm (1962), 317-330.

[2] G. Choquet, Mesures coniques, affines et cylindriques. Symposia Mathematica Vol. II 145-182. Acad. Press 1969.

[3] G. Choquet, Lectures on Analysis, Benjamin 1969.

[4] M. Hervé, Sur les representations intégrales à l'aide des points extremaux dans un ensemble compact convexe metrisable. C.R. Acad. Sci. (Paris) 253(1961), 336-368.

[5] A. Pietsch, Nuclear locally convex spaces, Ergebnisse der Mathematik, Band 66, Springer Verlag 1972.

[6] L. Schwartz, Radon Measures on Arbitrary Topological Spaces and Cylindrical measures, Oxford U.P. 1973.

[7] L. Schwartz, Sous-espaces hilbertiens d'espaces vectoriels topologiques et noyaux associés. J. Analyse Math. 13 (1964), 114-256.

Mathematisch Instituut

Postbus 800, Groningen

Netherlands

BOUNDEDNESS PROBLEMS FOR FINITELY ADDITIVE MEASURES

Philippe TURPIN

1. Introduction.

In this paper a "measure" is a finitely additive set function with values in a vector space. We shall consider measures defined on the σ-algebra P of all the subsets of the set $\mathbb{N}$ of non negative integers, but we could take for P an arbitrary σ-algebra.

If E is a t. v. space (topological vector space) a measure $\mu : P \to E$ is said to be bounded when $\mu(P) = \{\mu(H) \mid H \in P\}$ is bounded, i. e. absorbed (for non null homotheties) by every zero-neighbourhood.

$S(\mathbb{N})$ will be the normed subspace of $l^\infty(\mathbb{N})$ (with the induced norm) generated by the set of the characteristic functions 1_H, $H \in P$.

The following two theorems are known ([11] (prop. 0.5), [2]). They are other formulations of the fact that $S(\mathbb{N})$ is barrelled (cf. th. 3 below).

THEOREM 1. __If__ E __is a complete and metrizable locally convex t. v. space for a topology__ τ, __a measure__ $\mu : P \to E$ __is bounded for__ τ __if the convex hull__ $\mathrm{Conv}(\mu(P))$ __of__ $\mu(P)$ __is bounded for some Hausdorff linear topology__ ω __on__ E __coarser than__ τ.

THEOREM 1'(Uniform Boundedness Principle). __If__ E __is locally convex, if__ $(\mu_i)_{i\in I}$ __is a family of bounded measures__ $P \to E$ __and is pointwise bounded__ (__i. e.__ $\{\mu_i(H) \mid i \in I\}$ __is bounded in__ E __for every__ $H \in P$), __then__ $\{\mu_i(H) \mid i \in I,\ H \in P\}$ __is bounded.__

The object of this work is to examine the possibility of extending these theorems to non locally convex t. v. spaces, pose a few problems and give some counterexamples.

2. Problems.

2.1. An F-seminorm on a vector space E is a subadditive map $\nu : E \to R_+$ such that, for r scalar and $x \in E$, $\nu(rx) \to 0$ when $r \to 0$ and $\nu(rx) \leq \nu(x)$ if $|r| \leq 1$.

Problem 1. What t. v. spaces E verify the following property : for every pointwise bounded family of bounded measures $\mu_i : P \to E$, $i \in I$, we have for every continuous F-seminorm ν on E, $\sup\{\nu(\mu_i(H)) \mid i \in I, H \in P\} < \infty$.

Problem 2. For what complete metrizable t. v. spaces E is it true that $\sup\{\nu(\mu(H) \mid H \in P\} < \infty$ for every continuous F-seminorm ν on E and for every measure $\mu : P \to E$ which "behaves well" for some coarser linear topology on E ?

We shall see (th. 5 and 6) that some spaces do not verify such properties.

If, in problem 2, "behaves well" means "is σ-additive", N. J. Kalton has proved the following deep result ([4]).

THEOREM 2. If E is a separable complete metrizable topological group for a topology τ, a measure $\mu : P \to E$ is σ-additive if it is σ-additive for some Hausdorff group topology on E coarser than τ.

If "behaves well" means "Conv($\mu(P)$) bounded", I even do not know wether there exists some non locally convex space verifying the property of problem 2. And if this means simply "bounded" it would not be surprising that only finite dimensional spaces verify this property.

In particular, let us consider the case of locally p-convex spaces, $0 < p < 1$. A p-seminorm is an F-seminorm ν verifying $\nu(rx) = |r|^p \nu(x)$ for r scalar and a t. v. space is said to be locally p-convex when its topology can be defined by a family of p-seminorms. A t. v. space E is said to be p-barrelled when every l. s. c. (lower semi-continuous) p-seminorm on E is continuous.

THEOREM 3. If $0 < p \leq 1$, the following conditions are equivalent.

(i) $S(\mathbb{N})$ is p-barrelled

(ii) The property of theorem 1 is verified by every complete and metrizable locally p-convex space.

(iii) The property of theorem 1' is verified by every locally p-convex space.

These equivalences come from generalizations of the closed graph, Banach-Steinhaus and converse theorems ([16], pp. 10-21), a measure μ on P being extended to a linear map $x \to \int x\mu$ on $S(\mathbb{N})$. When the range space is locally p-convex, the boundedness of μ is equivalent to the continuity of its linear extension by a theorem of Rolewicz and Ryll-Nardzewski ([10], III, 6).

We have seen that the above conditions (i), (ii), (iii) hold when $p = 1$.

Problem 3. Is $S(\mathbb{N})$ p-barrelled for $0 < p < 1$?

2.2. A solution of these problems might use the notion of exhaustivity, which plays an important role in the case of locally convex spaces.

If E is a t. v. space and $\mathfrak{A}$ a ring of sets, a measure $\mu : \mathfrak{A} \to E$ is said to be exhaustive when $\mu(A_n)$ tends to 0 for every disjoint sequence (A_n) of $\mathfrak{A}$.

In theorem 2 above, σ-additivity may be replaced by exhaustivity: this was proved independently by N. J. Kalton and I. Labuda ([5], [6]).

If $\mu : \mathfrak{A} \to E$ is exhaustive, then $\sup\{\nu(\mu(A)) \mid A \in \mathfrak{A}\} < \infty$ for every continuous F-seminorm of E ([2]). This often implies the boundedness of μ ([3], [12] (6.6.5)), but not always ([13]).

We say that a linear topology ω on a vector space E is σ-exhaustive when every bounded measure $\mu : P \to (E, \omega)$ is exhaustive (then the same holds for an arbitrary σ-ring $\mathfrak{A}$ in place of P). These topologies are characterized in [8]: for example, every product of separable t. v. spaces is σ-exhaustive.

THEOREM 4. Let E be a t. v. space, with topology τ, verifying one of the following conditions (a), (b).

(a) There exists a linear σ-exhaustive topology ω on E which has the same bounded sets as τ.

(b) (E, τ) has a basis of zero-neighbourhoods closed for some linear σ-exhaustive topology ω on E coarser than τ.

Then, for every pointwise bounded set of bounded measures $\mu_i : P \to (E, \tau)$,

$i \in I$, <u>and for every τ-continuous F-seminorm</u> ν <u>on</u> E, $\sup\{\nu(\mu_i(H)) \mid i \in I,\ H \in P\}$ <u>is finite</u>.

If τ is itself σ-exhaustive this result is given by [2].

Let us prove the general case. By a lemma of [2] it is sufficient to prove that $\{\mu_i(H_n) \mid i \in I,\ n \in \mathbb{N}\}$ is bounded if (H_n) is a disjoint sequence of P.

If (a) is verified the μ_i's are exhaustive for ω, so for every sequence (α_n) extracted from $(\mu_i)_{i \in I}$ and for every null sequence $\varepsilon_n > 0$, $\varepsilon_n \alpha_n(H_n)$ tends to zero for ω by the Brooks-Jewett theorem (see [2]). This proves that $\{\mu_i(H_n) \mid i \in I,\ n \in \mathbb{N}\}$ is bounded for ω, and hence for τ.

Now assume that (b) is verified. Let V be a ω-closed τ-neighbourhood of 0 and let U be a balanced τ-neighbourhood of 0 verifying $U + U \subset V$. Let $S_{\mathbb{Z}}(\mathbb{N})$ be the subgroup of $S(\mathbb{N})$ consisting of sequences taking whole values, endowed with its "intrisic" topology λ ([5]). For every $h \in \mathbb{N}$, let C_h be the set of the elements x of $S_{\mathbb{Z}}(\mathbb{N})$ verifying $\int x\,\mu_i \in hU$ for every $i \in I$. $(C_h)_{h \in \mathbb{N}}$ is an increasing covering of $S_{\mathbb{Z}}(\mathbb{N})$. By [5] (th. 2), $\{1_{H_n} \mid n \in \mathbb{N}\}$ is included in the λ-closure of $C_h - C_h$ for some $h \in \mathbb{N}$. But, the μ_i's being exhaustive for ω, the mappings $x \mapsto \int x\,\mu_i$ from $S_{\mathbb{Z}}(\mathbb{N})$ to E are (λ, ω)-continuous ([5]). So, $\mu_i(H_n) \in hV$ for every $i \in I$ and every $n \in \mathbb{N}$.

<u>Problem 4</u>. What t. v. spaces (E, τ) verify the above property (a), or (b) ?

Locally convex spaces verify both these properties, taking for ω the weak topology. So theorem 4 gives the barrelledness of $S(\mathbb{N})$ and theorems 1 and 1'.

Other example: every (generalized) Orlicz space (or even Musielak-Orlicz space) $L^{\varphi}(T, \mathcal{A}, \alpha)$ on a measurable space $(T, \mathcal{A}, \alpha)$ verifies condition (b): when L^{φ} is not itself σ-exhaustive, take for ω the topology of convergence in measure on every set of finite measure, which by [9] is σ-exhaustive.

Theorem 6 below gives a space verifying neither (a) nor (b).

<u>Remark</u>. If (E, τ) possesses a countable basis $(B_n)_{n \in \mathbb{N}}$ of bounded sets (for example if it is locally bounded), condition (b) implies condition (a).

Indeed, assume that (b) is verified. We can take each B_n closed for ω, because in (E, τ) every bounded set is contained in a ω-closed bounded set, of the form $\bigcap_i r_i V_i$, where (V_i) is a basis of ω-closed τ-neighbourhoods of 0 and the r_i's are scalars. Then the finest linear topology ω_0 on E coinciding with ω on each B_n has the same bounded sets as τ ([12], 1.1.12) and is therefore σ-exhaustive, as ω. Observe that the couple (τ, ω_0) verifies both (a) and (b) ([12], 1.1.6).

2.3. paper [1] can give an idea of applications expected from solutions of the above problems. Let us give for example an application of the theorem 2 of Kalton to the Hardy classes H^p on the open unit disk D of the complex plane, improving a result of [1]. Let (z_n) be a sequence of D and consider the operator $u : H^p \to \mathbb{C}^{\mathbb{N}}$ defined by $u(f) = (f(z_n))_{n \in \mathbb{N}}$.

PROPOSITION 1. _If_ $0 < p \leq \infty$, $u(H^p)$ _contains_ l^∞ _if it contains_ 1_H _for every_ $H \subset \mathbb{N}$.

Proof. Endow $E = u(H^p)$ with the quotient topology τ from H^p. If μ is the measure $H \to 1_H$ defined on P and if $\mu(P) \subset E$, μ is σ-additive for the topology ω of pointwise convergence, which is coarser than τ. Then μ is bounded for τ, by th. 2 if $p < \infty$, by th. 1 if $p \geq 1$. The inclusion $S(\mathbb{N}) \subset E$ is therefore continuous, by the theorem of Rolewicz and Ryll-Nardzewski mentioned above ([10]) when $p < 1$. Whence $l^\infty \subset E$.

Proposition 1 is well known if $p = \infty$. It was proved in [1] for $p \geq 1$, using the barrelledness of $S(\mathbb{N})$. A positive answer to problem 3 would permit us to extend the method of [1] to the case $p > 0$.

3. Counterexamples.

Let $\varphi : R_+ \to R_+$ be continuous, increasing, subadditive and null at 0 and let A be a non void set. Then $l^\varphi(A)$ is the set of scalar families $x = (x_a)_{a \in A}$ verifying

$$|x|^\varphi = \sum_{a \in A} \varphi(|x_a|) < \infty .$$

Endowed with the F-norm $|.|^\varphi$, $l^\varphi(A)$ is a complete metrizable t. v. space, and $l^\varphi(A) \subset l^1(A)$ with continuous inclusion.

THEOREM 5. _If_ A _has the continuum power and if, for some real_ $p \in]0, 1[$, φ

verifies, when r tends to 0,

$$(1) \qquad \log^{-p}(|\log r|) = \mathcal{O}(\varphi(r))$$

then there exists a measure $\mu : P \to l^{\varphi}(A)$ which is σ-additive for some Hausdorff locally convex topology ω on $l^{\varphi}(A)$ coarser than the $|.|^{\varphi}$-topology (so $\mathrm{Conv}(\mu(P))$ is bounded for ω), but **verifies** $\sup_{H\in P} |\mu(H)|^{\varphi} = \infty$.

Proof. It suffices to construct a continuous injection $u : l^{\varphi}(A) \to l^{\infty}(\mathbb{N})$ such that $u(l^{\varphi}(A)) \supset S(\mathbb{N})$. Indeed, for such an u, we define a measure μ from P to $l^{\varphi}(A)$ by $\mu(H) = u^{-1}(1_H)$. μ is clearly σ-additive for the topology ω, inverse image by u of the topology $\sigma(l^{\infty}, l^1)$. But $\sup_{H\in P} |\mu(H)|^{\varphi} = \infty$. If not, $\mathrm{Conv}(\mu(P))$ would be bounded in $l^{\varphi}(A)$ by a theorem of Fischer and Schöler ([3], [7], [14]). By convexity, u^{-1} would then be continuous for the topology induced by l^{∞} and u would be an isomorphism of $l^{\varphi}(A)$ onto $l^{\infty}(\mathbb{N})$, which is obviously impossible.

Now we can assume that $A \subset P$ and that $\{1_a | a \in A\}$ is a Hamel basis for $S(\mathbb{N})$. Then we take for u the mapping defined by $u(x) = \sum_{a\in A} x_a 1_a$; this series converges in $l^{\infty}(\mathbb{N})$. (1) implies that u is injective. Indeed, suppose that $u(x) = 0$ with $x \neq 0 : 0 = \sum_0^{\infty} r_n 1_{a_n}$, with $|r_n|$ decreasing, $\sum \varphi(|r_n|) = s < \infty$, $r_0 \neq 0$ and $a_n \in A$, $a_n \neq a_m$ if $n \neq m$. For the norm $|.|^{\infty}$ of l^{∞} , the distance d_N of $r_0 1_{a_0}$ to $\sum_1^N r_n 1_{a_n}$ is less than $\sum_{n>N} |r_n| \leq \sum_{n>N} \varphi^{-1}(s/n)$, whence, with (1), $N^{1/p} = \mathcal{O}(\log|\log d_N|)$. But since 1_{a_0} is not generated by the 1_{a_n}'s, $1 \leq n \leq N$, d_N cannot decrease so quickly ([15], formula (2)).

Problem 5. For what Orlicz function φ does $l^{\varphi}(A)$ verify the property of the above theorem ?

$l^{\varphi}(A)$ cannot be locally convex (th. 1).

I do not know the answer for $l^p(A)$ with $0 < p < 1$.

The condition (1) can surely be weakened: it rests upon coarse metric evaluations in l_n^{∞} ([15]).

Let us now pass to a counterexample to the Uniform Boundedness Principle.

THEOREM 6. There exists an F-normed space $(E, \|.\|)$ and a pointwise bounded

<u>family of bounded measures</u> $\mu_i : P \to E$ <u>verifying</u> $\sup\{\|\mu_i(H)\| \mid i \in I, H \in P\} = \infty$.

Proof. Returning to the construction of u in the proof of th. 5, let us choose A so that $\{n\} \in A$ for every $n \in \mathbb{N}$. Then it is not difficult to show that, for every $M < \infty$, $\{1_H \mid H \in P\}$ is not included in the $|.|^\infty$-closure V_M of $\{y \in S(\mathbb{N}) \mid |u^{-1}(y)|^\varphi \leq M\}$ (see [15], lemma 2.5). But $V_M = \{y \in S(\mathbb{N}) \mid \nu(y) \leq M\}$ if ν is the greatest F-seminorm l. s. c. for $|.|^\infty$ and majorized by the F-norm $|u^{-1}(.)|^\varphi$. So, $\sup\{\nu(1_H) \mid H \in P\} = \infty$.

Now we can apply the following proposition, taking $S(\mathbb{N})$ for L and the set $\{1_H \mid H \in P\}$ for X. This proposition is a modification of a theorem due to L. Waelbroeck ([16], pp. 10-13).

PROPOSITION 2. <u>For every subset</u> X <u>of a</u> t. v. <u>space</u> L, <u>the following conditions are equivalent</u>.

(i) $\sup\{\nu(x) \mid x \in X\} < \infty$ <u>for every l. s. c. F-seminorm</u> ν <u>on</u> L.

(ii) <u>For every t. v. space</u> F, <u>for every pointwise bounded family of continuous linear mappings</u> $u_i : L \to F$, $i \in I$, <u>and for every continuous F-seminorm</u> ρ <u>on</u> F, $\sup\{\rho(u_i(x)) \mid i \in I, x \in X\} < \infty$.

The implication (i) $\Rightarrow$ (ii) is easily established.

The proof of the converse (adapted from [16]) uses the following fact.

If ν is a l. s. c. F-seminorm on L, there exists a vector space F, an F-seminorm ρ on F and a pointwise bounded family of linear maps $u_i : L \to F$ continuous for ρ and verifying $\nu(x) = \sup_{i\in I} \rho(u_i(x))$ for every $x \in L$.

Indeed, let $\mathcal{R}$ be a set of F-seminorms defining the topology of L and let $I = \{2^h r \mid h \in \mathbb{N}, r \in \mathcal{R}\}$. F is the direct sum $L^{(I)}$, ρ is defined by

$$\rho(x) = \sup_{i\in I} \inf\{\nu(a) + i(b) \mid a \in L, b \in L, x_i = a + b\}, \quad x \in F,$$

u_i is the canonical ith injection of L into F. If $\lambda > \rho(u_i(x))$ for every i, then, for every $r \in \mathcal{R}$ and $h \in \mathbb{N}$, $x = a + b$ with $\nu(a) < \lambda$ and $r(b) < 2^{-h}\lambda$, whence $\nu(x) \leq \lambda$ since ν is l. s. c.. So, $\nu(x) \leq \sup_i \rho(u_i(x))$. The rest is easily verified.

References.

[1] BENNETT (G), KALTON (N. J.), Inclusion theorems for K-spaces, Canad. J. Math. 25 (1973), 511-524.

[2] DREWNOWSKI (L), Uniform Boundedness Principle for finitely additive vector measures, Bull. Acad. Polon. Sci. 21 (1973), 115-118.

[3] FISCHER (W), SCHÖLER (U), The range of vector measures into Orlicz spaces, Studia Math. 59 (1976), 53-61.

[4] KALTON (N. J), Subseries convergence in topological groups and vector spaces, Israel J. Math. 10 (1971), 402-412.

[5] ——, Topologies on Riesz groups and applications to measure theory, Proc. London Math. Soc. (3) 28 (1974) 253-273.

[6] LABUDA (I), A generalization of Kalton's theorem, Bull. Acad. Polon. Sci. 21 (1973) 509-510.

[7] ——, Ensembles convexes dans les espaces d'Orlicz, C. R. Acad. Sci. Paris 281A (1975), 443-445.

[8] ——, Sur les mesures exhaustives et certaines classes d'espaces vectoriels topologiques considérés par W. Orlicz et L. Schwartz, C. R. Acad. Sci. Paris 280A (1975) 997-999.

[9] MATUSZEWSKA (W), ORLICZ (W), A note on modular spaces IX, Bull. Acad. Polon. Sci. 16 (1968), 801-808.

[10] ROLEWICZ (S), Metric Linear Spaces, PWN, Warsaw 1972.

[11] THOMAS (G. E. F), The Lebesgue-Nikodym theorem for vector valued Radon measures, Mem. Amer. Math. Soc. 139 (1974).

[12] TURPIN (Ph.), Convexités dans les espaces vectoriels topologiques généraux, Dissertationes Math. 131.

[13] ——, Une mesure vectorielle non bornée, C. R. Acad. Sci. Paris 280A (1975) 509.

[14] ——, Colloque sur l'Intégration vectorielle et multivoque, Caen 22 et 23 mai 1975, OFFILIB, Paris.

[15] ——, Properties of Orlicz-Pettis or Nikodym type and barrelledness conditions, Ann. Inst. Fourier (to appear).

[16] WAELBROECK (L), Topological Vector Spaces and Algebras, Lecture Notes 230, Springer, Berlin 1971.

Univ. de Paris XI,
Centre d'Orsay,
Mathematiques (Bat. 425),
91405 Orsay, France.

VECTOR MEASURES AND THE ITO INTEGRAL

by John B. Walsh

It seems appropriate at a conference on vector-valued measures to look at one of the best examples of one: the stochastic integral. It is not obvious from the usual construction that the stochastic integral is at all connected with vector measures. It looks more like a form of Riemann integral, but as such it has some puzzling properties. For instance, in the approximating Riemann sums, the integrand must be evaluated at the left-hand endpoints of the intervals. Why not in the middle? Why not at the right? These are natural questions, but one feels that they are of the kind that would not come up if it were really a Riemann integral in any reasonable sense. Once we see that there is in fact a vector-valued measure and that the Ito integral is really an integral in the usual sense with respect to this measure, they will practically disappear.

The approach we will describe below is, except for some minor modifications, due to J. Pellaumail. Since the detailed proofs are carefully set out in [2] and [4], we will be more concerned with showing informally why things ought to be true than with proving that they actually are true. We will also confine ourselves to the most elementary part of the theory; the reader will find further - and deeper - developments in [2] and [4].

1. ITO MEASURE.

Let $\{B_t, t \geq 0\}$ be a standard Brownian motion. We want to define

$$\int_0^t \phi_s \, d B_s \tag{1}$$

for suitable integrands ϕ. The first difficulty is that $t \to B_t$ is not of bounded variation - it is even nowhere differentiable - and there is no way to write it as the difference of two increasing functions. Thus we can't define (1) as a Stieltjes integral. In fact, consider the famous integral

$$\int_0^t B_s \, d B_s = \tfrac{1}{2} B_t^2 - \tfrac{1}{2} t . \tag{2}$$

If this were a Stieltjes integral, we would not get the $\frac{1}{2}t$ term. But let's look at it from the point of view of a vector integral.

If we want to write (1) and (2) as integrals with respect to a vector measure B, we first have to find the measure. To do this we need to answer the following two questions.

(i) On what σ-field is B defined?

(ii) In which space does it take its values?

To answer (i), let's consider which sets we can measure. These certainly include the half-open intervals of the form $(s, t]$ - think of B_t as the distribution function of the measure B - for the B-measure of $(s, t]$ should be $B_t - B_s$. But this is not enough. We will need to integrate random functions, as in (2) for instance, so we will need to be able to measure random intervals. Let $S \leq T$ be bounded stopping times and let

$$(S, T] = \{(t, \omega) : S(\omega) < t \leq T(\omega)\}$$

be a stochastic interval. Then we can put $B(S, T] = B_T - B_S$. Notice that a stochastic interval is a subset of $\mathbb{R}_+ \times \Omega$, not of $\mathbb{R}_+$, so our σ-field will be composed of subsets of $\mathbb{R}_+ \times \Omega$, not of the line. The simplest σ-field to take is surely the σ-field $\mathcal{P}$ generated by the stochastic intervals of the form $(S, T]$, where S and T are bounded stopping times, which is called the σ-field of <u>predictable sets</u>. It has the property that any process $X = \{X_t, t \geq 0\}$ which is adapted to $(\mathcal{F}_t)$ and left continuous as a function of t is $\mathcal{P}$-measurable. We say that a $\mathcal{P}$-measurable process is <u>predictable</u>. Thus B_t itself is predictable so that, for instance, $\mathcal{P}$ is large enough to allow us to define the integral (2). (We should underline the fact that a process X is really a function of two variables, $X_t(\omega)$, even though the ω is usually suppressed, so that X is predictable iff it is $\mathcal{P}$-measurable as a function of the two variables).

Now that we have decided that B will be a measure on $\mathcal{P}$, let's see what vector space is involved. It turns out that there are many possibilities, and the choice depends somewhat on the use one wants to make of the integral, and somewhat on his taste. Pellaumail and Métivier use $L^p(\Omega, \mathcal{F}, P)$ and $L^0(\Omega, \mathcal{F}, P)$, this last being the space of all random variables, given the topology of convergence in measure. These are probably the best spaces to take in most cases, but in order to understand the essence of the stochastic integral, we would like to take a vector space which expresses its most important properties. We have two properties in mind. The first is that $t \to \int_0^t \phi_s \, dB_s$ should be continuous. Indeed, Itô first introduced the stochastic integral in order to solve stochastic differential equations. The solutions of these are diffusions, which model physical particles which diffuse through a liquid and whose motions are therefore continuous. It would have been embarrassing had the stochastic integrals turned out to be discontinuous!

The second property is that the process $\{\int_0^t \phi_s \, dB_s, t \geq 0\}$ is a martingale. One has to work with stochastic integrals to appreciate how vital this is. It is one of the main reasons that they can be defined for such a large class of integrands, and it comes up continually in applications.

Let's consider the square-integrable case, which is always the easiest to handle in this theory. Let $\mathcal{M}^2$ be the set of all right-continuous martingales

$\{M_t, \mathcal{F}_t, t \geq 0\}$ such that $M_0 \equiv 0$, and such that $||M^2|| \overset{\text{def}}{=} \sup_t E\{M_t^2\}$ is finite. We let $\mathcal{M}_c^2$ be the subset of all continuous martingales in $\mathcal{M}^2$. (Since the $\mathcal{F}_t$ are the fields generated by Brownian motion, it turns out that $\mathcal{M}_c^2 = \mathcal{M}^2$, but this is a rather special property of Brownian motion and we don't want to use it here).

By the martingale convergence theorem, each $M \in \mathcal{M}^2$ has a limit, M_∞, and $M_t = E\{M_\infty | \mathcal{F}_t\}$. Furthermore, Doob's maximal inequality tells us that

$$E\{(\sup_t M_t)^2\} \leq 4E\{M_\infty^2\} = 4||M||^2.$$

This tells us that if (M^n) is a sequence in $\mathcal{M}^2$, then $||M^n - M|| \to 0$ implies that $\sup_t |M_t^n - M_t| \to 0$ in L^2. It follows that $\mathcal{M}_c^2$ is a closed subspace of $\mathcal{M}^2$. In fact, we can always identify a martingale in $\mathcal{M}^2$ with its limit, for if M_∞ is any square-integrable random variable, we can define a martingale $M \in \mathcal{M}^2$ by $M_t = E\{M_\infty | \mathcal{F}_t\}$. This correspondence $M \to M_\infty$ is therefore an isomorphism between $\mathcal{M}^2$ and $L^2(\Omega, \mathcal{F}, P)$, where we of course identify martingales which are a.s. equal. An immediate consequence is that $\mathcal{M}^2$ and $\mathcal{M}_c^2$ are both Hilbert spaces.

With this established, let's define our measure B. Let $\Lambda = (0, S]$, where S is a bounded stopping time, and put

$$B(\Lambda) = \{B_{S\wedge t}, t \geq 0\}.$$

Note that $B(\Lambda)$ is actually a stochastic process. It is a continuous martingale by the stopping theorem, and moreover, it is square-integrable: just apply the stopping theorem to the martingale $B_t^2 - t$ at the time S to see that $E\{B_S^2\} = E\{S\}$. Thus $B(\Lambda) \in \mathcal{M}_c^2$, and moreover

$$||B(\Lambda)||^2 = E\{S\} = \int_\Lambda dt \times dP.$$

The last equality is significant. It tells us that if ν is the measure $d\nu = dt \times dP$ on $\mathbb{R}_+^2 \times \Omega$, then

(3) $$||B(\Lambda)||^2 = \nu(\Lambda).$$

Thus ν is a real-valued measure on $\mathcal{B} \times \mathcal{F}$, where $\mathcal{B}$ is the Borel field of $\mathbb{R}_+^2$, which controls B. We will refer to it as the <u>controlling measure</u>.

To extend B to $\mathcal{P}$, we first notice that if $S \leq T$ are bounded stopping times, we can set

$$B(S, T] = B(0, T] - B(0, S] = \{B_{T\wedge t} - B_{S\wedge t}, t \geq 0\}.$$

We then define B on the class $\mathcal{S}$ of all finite unions of stochastic intervals in the obvious way, and verify that (3) holds for each $\Lambda \in \mathcal{S}$. But $\mathcal{S}$ is a field which generates $\mathcal{P}$. Apply the usual approximation theorem for real-valued measures to ν: if $\Lambda \in \mathcal{P}$, and Λ $[0, t] \times \Omega$, some t, there exists a sequence $\Lambda_n \subset \mathcal{S}$ such that $\nu(\Lambda_n \Delta \Lambda) \to 0$. From (3), $\{B(\Lambda_n)\}$ is Cauchy in $\mathcal{M}^2$, and we just put $B(\Lambda) =$

$\lim B(\Lambda_n)$. It is easy to see that (3) still holds, so that the countable additivity of B on $\mathcal{P}$ is nearly automatic. Indeed, if $\Lambda_n \subset [0, t] \times \Omega$ is a decreasing sequence in $\mathcal{P}$ with $\bigcap_n \Lambda_n = \phi$, then $\lim_n ||B(\Lambda_n)||^2 = \lim_n \nu(\Lambda_n) = 0$.

Since $B(\Lambda_n) \in \mathcal{M}_c^2$ and $\mathcal{M}_c^2$ is closed, $B(\Lambda)$ is also in $\mathcal{M}_c^2$, so B is a measure on $\mathcal{P}$ with values in $\mathcal{M}_c^2$. The reason for specifying that $\Lambda \subset [0, t] \times \Omega$ above is simply to insure that $||B|| < \infty$; B is only σ-finite, not finite.

Even though the above measure was introduced by Pellaumail, it seems just to call it Ito measure, since it gives the Ito integral. To resume what we have established above:

Theorem 1 There exists a σ-finite measure B on the predictable sets $\mathcal{P}$, called Ito measure, which has values in $\mathcal{M}_c^2$, and is such that

(i) if T is a bounded stopping time, $B(0, T] = \{B_{T\wedge t}, 0 \leq t < \infty\}$;

(ii) if $\Lambda \in \mathcal{P}$ and $\nu(\Lambda) < \infty$, $||B(\Lambda)||^2 = \nu(\Lambda)$.

2. THE ITO INTEGRAL.

The existence of the controlling measure ν makes integration with respect to B very simple, and there is no need to call on any of the deeper theories of integration with respect to a vector measure; the integral can be constructed directly. Let's consider this done, and check that this integral agrees with the Ito integral. This is a matter of following the usual construction step-by-step.

First, if $S \leq T$ are bounded stopping times and if Y is a square-integrable $\mathcal{F}_S$-measurable random variable, let $\phi_t(\omega) = Y(\omega) \; I_{(S,T]}(t)$. Note that ϕ is $\mathcal{P}$-measurable, for it is left-continuous in t and, since $Y \in \mathcal{F}_S$, it is also adapted. Thus, as a vector integral

$$\int \phi dB = Y\, B(S,T] = \{Y(B_{T\wedge t} - B_{S\wedge t}),\; t \geq 0\}.$$

This agrees with Ito's definition. Notice that

$$(4) \qquad ||\int \phi dB||^2 = \int_{\mathbb{R}_+ \times \Omega} \phi_t(\omega) \quad d\nu(t, \omega).$$

If ϕ is a finite sum of such functions, we call it a simple function. The integral extends to simple functions by linearity, and (4) continues to hold. If ϕ is now $\mathcal{P}$-measurable, such that $\int_{\mathbb{R}_+ \times \Omega} \phi_t^2(\omega) \; d\nu(t, \omega) < \infty$ - in other words, if $\phi \in L^2(\mathbb{R}_+ \times \Omega, \mathcal{P}, \nu)$ - Ito shows there are simple ϕ^n such that $\int |\phi^n - \phi|^2 \, d\nu \to 0$, and he defines $\int \phi \, dB$ to be the limit of $\int \phi^n \, dB$. But, by (4), $(\int \phi^n \, dB)$ forms a Cauchy sequence in $\mathcal{M}_c^2$, and its limit is none other than $\int \phi dB$, calculated this time as an integral with respect to the vector measure B. Once more, (4) holds. This gives us

Theorem 2. Let ϕ_t be a predictable process such that $\int_{IR_+^2 \times \Omega} \phi_t^2(\omega)\, d\nu(t, \omega) < \infty$. Then

(i) $\int \phi\, dB$ exists in $\mathcal{M}_c^2$;

(ii) $\| \int \phi\, dB \|^2 = \int \phi^2 d\nu$;

(iii) $\int \phi\, dB$ agrees with the Ito integral.

We might remark in passing that (i) and (ii) imply that the stochastic integral is a norm-preserving mapping from $L^2(IR_+ \times \Omega, \mathcal{P}, \nu)$ into $\mathcal{M}_c^2$. It is this property that makes the theory of stochastic integration so pleasant in the square-integrable case.

3. SOME GENERALIZATIONS.

Pleasant as the theory of stochastic integration is in the square-integrable case, it is not sufficient for many applications. We have seen how to integrate predictable ϕ for which $E\{\int_0^\infty \phi_s^2\, ds\} < \infty$, but in fact, the proper condition on ϕ is not this, but that $\int_0^t \phi_s^2\, ds < \infty$ a.s. for all $t > 0$. This is a much handier condition. For instance, it is always satisfied if $t \to \phi_t$ is continuous - which is the case if $\phi_t = f(B_t)$ for a continuous f - since ϕ is then bounded on $[0,t]$. On the other hand, if we begin integrating such ϕ, we will quickly leave $\mathcal{M}_c^2$, and we will need to go to a larger space. A reasonable space to choose would be a space of local martingales. A local martingale $\{M_t, \mathcal{F}_t, t \geq 0\}$ is a process with the property that there exists a sequence (T_n) of stopping times increasing to ∞ such that for each n the process $\{I_{\{T_n > 0\}} M_{T_n \wedge t}, \mathcal{F}_t, t \geq 0\}$ is a martingale. (The reason for $I_{\{T_n > 0\}}$ in the definition is simply that M_0 might not be integrable, and we could then set $T_n = 0$ on the set $\{|M_0| < n\}$, which would assure us that $I_{\{T_n > 0\}} M_0$ is integrable).

Let $\mathcal{M}^0$ be the class of all right-continuous local martingales $\{M_t, \mathcal{F}_t, t \geq 0\}$, $\mathcal{F}_t$ always being the fields generated by Brownian motion, such that $M_0 = 0$. We provide $\mathcal{M}^0$ with the topology of uniform convergence in probability on compact intervals:

$$M^n \to 0 \text{ in } \mathcal{M}^0 \quad \text{iff} \quad \sup_{s \leq t} |M_s| \to 0 \text{ in probability for all } t > 0.$$

This is a metrizable topology; a compatible distance is

$$d(M, N) = \sum_n 2^{-n} E\{\sup_{t \leq n} \frac{|M_t - N_t|}{1 + |M_t - N_t|}\}.$$

$\mathcal{M}^0$ has the inconvenience of being incomplete, but one can show that the subspace $\mathcal{M}_c^o$ of continuous local martingales is complete, and closed in $\mathcal{M}^o$. Since $\mathcal{M}_c^o$ is all we will use, the incompleteness of $\mathcal{M}^o$ won't affect us.

Note that $\mathcal{M}^2$ can be embedded in $\mathcal{M}^o$, and that $M^n \to M$ in $\mathcal{M}^2$ implies $M^n \to M$ in $\mathcal{M}^o$. Conversely if M^n is a sequence in $\mathcal{M}^2$ such that $\{(M^n)^2\}$ is uniformly integrable - which is slightly stronger than saying the M^n are norm-bounded in $\mathcal{M}^2$ - then $M^n \to M$ in $\mathcal{M}^2$ iff $M^n \to M$ in $\mathcal{M}^o$. Thus the Ito measure B can also be regarded as a measure with values in $\mathcal{M}^o$, with the additional advantage that it is now finite, not just σ-finite. Indeed, $B\{\mathbb{R}_+ \times \Omega\} = \{B_t, t \geq 0\}$ which is certainly a local martingale - even a martingale.

It is now easy to define $\int \phi dB$ for a predictable ϕ such that $\int_o^t \phi_s\, ds$ for all $t > 0$. Just let T_n be the stopping time

$$T_n = \inf\{t: \int_o^t \phi_s^2\, ds \geq n\},$$

and then set $\phi_t^n = \phi_t\, I_{\{t \leq T_n\}}$. It is not difficult to see that ϕ_t^n is predictable, and clearly

$$\int_{\mathbb{R}_+^2 \times \Omega} \phi^n\, d\nu = E\{\int_o^\infty (\phi_s^n)^2\, ds\} \leq n.$$

Thus $\int \phi^n\, dB$ exists in $\mathcal{M}^2$, hence in $\mathcal{M}^o$. Furthermore, it is easy to see that $(\int \phi^n\, dB)(t) = (\int \phi^{n+1}\, dB)(t)$ on the set $\{t < T_n\}$, since $\phi_s^n \equiv \phi_s^{n+1}$ on $[0, T_n]$. Thus we can define

$$\int \phi\, dB = \lim_{n \to \infty} \int \phi^n\, dB,$$

where the limit exists in $\mathcal{M}^o$ because $T_n \to \infty$, and the limit is indeed a local martingale: when stopped at T_n, it is even in $\mathcal{M}_c^2$.

We have only discussed stochastic integration with respect to Brownian motion, but it is possible to define integration with respect to any martingale $M \in \mathcal{M}^2$ just as we did above, with two differences. If M is discontinuous, the integrals will be right continuous, but not necessarily continuous, and the controlling measure will no longer be $dt \times dP$, but

$$\nu_M(\Lambda) = E\{\int I_\Lambda(s, \omega)\, d\langle M\rangle_s\},$$

where $\langle M\rangle$ is the unique predictable increasing process such that $\langle M\rangle_o = 0$ and $M_t^2 - \langle M\rangle_t$ is a martingale.

One can go beyond this: under proper conditions on the predictable process ϕ, one can define $\int \phi\, dX$ in case X is local martingale, or more generally, a semi-martingale, i.e. a process of the form

$$X_t = M_t + V_t,$$

where M is a right-continuous local martingale and V is an adapted process of locally bounded variation, that is, the difference of two increasing processes.

In this case, one merely defines

$$(4) \qquad \int\phi\, dX = \int\phi\, dM + \int\phi dV,$$

where the last term is a Lebesgue - Stieltjes integral. This is no longer a local martingale because of the last term, but as the first term on the right-hand side of (4) is a local martingale, while the second is a process of locally bounded variation, the integral itself will be a semi-martingale. We could thus write (4) as an integral with respect to a measure with values in a suitable space of semi-martingales. Some vector spaces of semi-martingales have been constructed in (1), but this involves a certain amount of probabilistic machinery which we don't have at hand, so we won't go into it here.

The step from integration with respect to local martingales to that with respect to semi-martingales is a small one, but, according to a remarkable theorem of Métivier, it is the last step: stochastic integration with respect to semi-martingales is as far as one can go. To see this, we must consider measures with values, not in a space of semi-martingales (which would make the result circular) but in the space $L^0(\Omega, \mathcal{F}, P)$ of random variables, given the metric of convergence in probability.

Let $\{X_t, t \geq 0\}$ be an adapted stochastic process. Suppose X generates a map m of $\mathcal{P}$ into L^0, such that if $S \leq T$ are stopping times, $m(S, T] = X_T - X_S$. We suppose m is σ-finite, in the sense that there exists a sequence of stopping times (T_n) increasing to infinity such that

$$(5) \qquad E\{|m(\Lambda)|\} < \infty \text{ if } \Lambda \in \mathcal{P} \text{ and } \Lambda \subset [0, T_n].$$

Notice that m must be local in the sense that if $S \leq T \leq T_n$ are stopping times, and if $F \in \mathcal{F}_S$, then

$$(6) \qquad m\{(S, T] \cap \mathbb{R}_+ \times F\} = I_F m\{(S, T]\},$$

for, if we set $S' = S$ on F and $S' = T$ on F^c, then S' is a stopping time, and both sides of (6) equal $m\{(S', T]\} = X_T - X_{S'}$. Now consider m without reference to X. If m is σ-additive, and if m satisfies (5) and (6), we say is a <u>local stochastic measure</u>. We then say m is <u>generated by</u> X if $m\{(S, T]\} = X_T - X_S$. Clearly, any reasonable stochastic integral should give rise to a stochastic measure. Métivier then proves:

<u>Theorem 3</u> (Métivier) Every local stochastic measure is generated by a semi-martingale.

We should mention that Métivier treats a much more general case then we do here, allowing both the integrand and the process X to take values in Banach spaces.

4. SOME LOOSE ENDS.

The main contribution of the vector approach, pedagogical advantages apart, is probably the added insight it gives into stochastic integration, and above all the possibility it offers of posing questions that would be difficult to state in another setting. Métivier's theorem is an example of this; it is hard to see how one could have posed the problem in the classical theory.

Let's look at the role played by the predictable σ-field from this viewpoint, Returning to integration with respect to Brownian motion, recall that we know how to integrate predictable ϕ. It follows that we can integrate ϕ which are measurable with respect to the completion of $\mathcal{P}$. "Completion" means of course the completion with respect to Ito measure, but since we have a real-valued measure ν with the same null-sets as B, this completion is the same as the completion $\mathcal{P}^{\nu}$ of $\mathcal{P}$ with respect to ν.

Now Ito actually defined the stochastic integral for adapted measurable ϕ rather than for predictable ϕ, i.e. for ϕ which are measurable with respect to $\mathcal{A}$, where

$$\mathcal{A} = \{\Lambda \subset \mathbb{R}_+ \times \Omega : \Lambda \in \mathcal{B} \times \mathcal{F} \text{ and } I_\Lambda(t,.) \text{ is } \mathcal{F}_t \text{ - measurable}, \forall t\}.$$

But it is not difficult to show that $\mathcal{A}$ is a sub-σ-field of $\mathcal{P}^{\nu}$ - this was actually proved by Ito, though not in those terms - and the same is true if B_t is replaced by any continuous martingale. However, in general the natural integrands are the predictable processes and only the predictable processes. Here is an example to illustrate this.

Let X_t be the Poisson process with $X_o = 0$ and let be the first jump time of X. Recall that τ is an exponential random variable with $E\{\tau\} = 1$. We set $M_t = X_t - t$. M_t is a martingale relative to the natural σ-fields $\mathcal{F}_t$ generated by X. These fields have one particularity we will need: the set $\{t < \tau\}$ is an atom of $\mathcal{F}_t$. (If $t < \tau$, X is identically zero on $[0, t]$, so how could it be otherwise?) Consequently, any stopping time S must be constant on the set $\{S < \tau\}$ We can construct the Ito measure M for M_t exactly as we did for Brownian motion. Once again, M is controlled by the real-valued measure $d\nu = dt \times dP$, i.e. if $\Lambda \in \mathcal{P}$, $||M(\Lambda)||^2 = \nu(\Lambda)$. (This follows as before from the fact that $M_t^2 - t$ is a martingale). We claim that $\mathcal{A}$ is <u>not</u> in the ν-completion $\mathcal{P}^{\nu}$ of $\mathcal{P}$.

One set which is in $\mathcal{A}$ but not in $\mathcal{P}$ is the set $[\tau]$, the graph of τ :

$$|\tau| = \{(t, \omega) : t = \tau(\omega)\}.$$

This set is even well-measurable, so it is certainly in $\mathcal{A}$. To show it is not in $\mathcal{P}^{\nu}$, we will show that its inner measure is zero and its outer measure one (which would suggest that not all well-measurable sets are measurable!)

It is clear that the inner measure of $[\tau]$ is zero, so let's compute its outer measure, which is

$$\inf \left[\nu\{\bigcup_n (S_n, T_n]\} \right],$$

where the infimum is over all sequences $(S_n, T_n]$ of stochastic intervals whose union contains $[\tau]$. Notice that we can replace S_n and T_n in the above by $S_n \wedge \tau$ and $T_n \wedge \tau$ respectively without increasing the ν-measure of the union, and the union will still contain $[\tau]$. Moreover, as S_n and T_n are constant on $\{S_n < \tau\}$ and $\{T_n < \tau\}$respectively, there are real s_n and t_n for which $S_n \wedge \tau = s_n \wedge \tau$ and $T_n \wedge \tau = t_n \wedge \tau$. If $\Lambda = \bigcup_n (s_n \wedge \tau, t_n \wedge \tau]$ and $A = \bigcup (s_n, t_n]$, then $\Lambda = (0, \tau] \wedge (A \times \Omega)$. Since $[\tau] \subset \Lambda$, $P\{\tau \in A\} = 1$ and $P\{\tau \in \mathbb{R}_+ - A\} = 0$. But τ has an exponential distribution, so $\mathbb{R}_+ - A$ must have zero Lebesgue measure, hence $\nu\{(\mathbb{R}_+ - A) \times \Omega\} = 0$. Thus

$$\nu(\Lambda) = \nu\{(0,\tau]\} \wedge A \times \Omega\} = \nu\{(0, \tau]\} = E\{\tau\} = 1.$$

This shows that $[\tau]$ has outer ν-measure 1, and we are done.

We should point out that this does not mean that it is impossible to define $M[\tau]$. It just shows that $M[\tau]$ can not be determined by the values of M on $\mathcal{P}$. In fact it is possible to define it consistently and, more generally, to extend M to the well-measurable sets, but one needs to use some new principles to do this (see [3]).

To close this article, let's return to the question of approximating Riemann sums raised in the introduction. We will answer it by calculating $\int_0^t B_s dB_s$. To do this, we must approximate the integrand, B_t, by simple functions ϕ_n. Suppose ϕ_n is constant on each interval $(\frac{k}{n}, \frac{k+1}{n}]$, taking on the value $B_{\frac{k}{n}}$. Then

$$\phi_n(t) = \sum_{k=0}^{n-1} B_{\frac{k}{n}} \, I_{(\frac{k}{n}, \frac{k+1}{n}]}(t) \tag{7}$$

Note that ϕ_n is adapted and left-continuous, hence predictable. Let B be Ito measure; for simplicity we will take its values in L^2, rather than in $\mathcal{M}^2$. Then ϕ_n is B-integrable, and it is easily checked that $\int_{[0,1] \times \Omega} (\phi_n - B)^2 d\nu \to 0$, so that $\int \phi_n \, dB \to \int B \, dB$ in L^2. But now

$$\int \phi_n \, dB = \sum_{k=0}^{n-1} B_{\frac{k}{n}} \, B\{(\tfrac{k}{n}, \tfrac{k+1}{n}] \times \Omega\} \tag{8}$$

which is just

$$= \sum_{k=0}^{n-1} B_{\frac{k}{n}} \, (B_{\frac{k+1}{n}} - B_{\frac{k}{n}}). \tag{9}$$

This last is not a Riemann sum, but it certainly looks like one. As promised, the integrand has been evaluated at the left-hand end point of each interval $(\frac{k}{n}, \frac{k+1}{n}]$.

Why not at the center or the right? This last would correspond to replacing ϕ_n by

$$\hat{\phi}_n(t) = \sum_{k=0}^{n-1} B_{\frac{k+1}{n}} I_{(\frac{k}{n}, \frac{k+1}{n}]}(t) .$$

But $\hat{\phi}_n$ is not predictable - in fact if $t < 1$, $\hat{\phi}_n(t)$ is not even $\mathcal{F}_t$- measurable - so that $\int \hat{\phi}_n \, dB$ is not defined, and, even if we could give it a sense - and we could, of course - there is no guarantee that it would converge to $\int BdB$. (In fact, it does not - see below).

So the reason that we can not evaluate the integrand on the center or the right of the interval is simply that this is tantamount to approximating the integrand by a non-measurable function.

This is the point we wanted to make, but let's complete the calculation for form's sake. We can rewrite (9) as

$$\frac{1}{2} \sum_{k=0}^{n-1} (B^2_{\frac{k+1}{n}} - B^2_{\frac{k}{n}})^2) = \frac{1}{2} B_1^2 - \frac{1}{2} \sum_{k=0}^{n-1} (B_{\frac{k+1}{n}} - B_{\frac{k}{n}})^2 .$$

But this last sum is famous - in certain circles at least - since it converges to the quadratic variation of Brownian motion on $[0, 1]$ as $n \to \infty$, and this quadratic variation is, by an elementary but celebrated result, almost surely exactly equal to one. Thus

$$\int_0^1 B \, dB = \frac{1}{2} B_1^2 - \frac{1}{2}$$

It is amusing to repeat this calculation, evaluating the integrand in (9) first at the center, then at the right of the interval. In place of $\frac{1}{2} B_1^2 - \frac{1}{2}$, one gets $\frac{1}{2} B_1^2$ and $\frac{1}{2} B_1^2 + \frac{1}{2}$, respectively.

REFERENCES

(1) M. EMERY: Stabilité des solutions des equations differentielles stochastiques; applications aux intégrales multiplicatives stochastiques. (To appear).

(2) M. METIVIER: The stochastic integral with respect to processes with values in a reflexive Banach space. Theory of Prob. and its Appl. 29 (1974), pp. 758-787.

(3) P-A. MEYER: Un cours sur les intégrals stochastiques; Seminaire de Probabilité de l'Université de Strasbourg X, Lecture Notes in Math. 511, Springer-Verlag, Berlin, 1976.

(4) J. PELLAUMAIL: Une nouvelle construction de l'intégrale stochastique; Asterisque 9.

John B. Walsh, Department of Mathematics, University of British Columbia, Vancouver, B.C. Canada.

INFINITELY DIVISIBLE STOCHASTIC DIFFERENTIAL EQUATIONS IN SPACE-TIME

Aubrey Wulfsohn
The Open University
Milton Keynes
Buckinghamshire
U.K.

ABSTRACT

We consider 'stochastic differential equations' of the symbolic form

$$\frac{\partial n}{\partial t} = \alpha(n) + \sigma(n) \frac{\partial \beta}{\partial t}$$

where n denotes a random process in space-time, α and σ are measurable functions and $\frac{\partial \beta}{\partial t}$ is a spatially uncorrelated space-time white noise. For each t we take n_t to be a random variable valued measure. Defining a suitable topology for random measures we approximate given initial conditions by those for which the solution is a sum of independent random processes. The 'vague' solution of the equation will be a limit of these approximating sums; when it exists it is infinitely divisible.

1. Introduction

We wish to deal with symbolic stochastic differential equations of the type encountered for geographically structured population processes and continuous branching diffusion processes, where each individual wanders according to a Markov process diffusion process with generator A. Work on these has been initiated in [1]. The equations are of the form

$$\frac{\partial n}{\partial t} = \alpha(n) + An + \sigma(n) \frac{\partial \beta}{\beta t}$$

where we allow α and σ to vary with geographic position. The equation may be written

$$\frac{\partial}{\partial t} n_t(x) = \alpha_x(n_t(x)) + An_t(x) + \sigma_x(n_t(x)) \frac{\partial}{\partial t} \beta_t(x)$$

Here $n_t(x)$ is a random variable and represents the population density at time t at a place x in a region $X \subset R^d$, so we assume $n_t(x) \geq 0$; for fixed x both α_x and σ_x are measurable real-valued functions: $\beta_t(x)$ on T×X, T = $[0,\infty)$, is a space-time Brownian motion, i.e. the increments $\Delta_s \beta_t = \beta_{t+s} - \beta_t$ are Gaussian random variables with zero mean, and, symbolically, covariance

$$E(\Delta_s \beta_t(x), \Delta_x \beta_t(y)) = \delta(x-y)s + o(s)$$

where δ is the Dirac delta 'function'. For fixed $x \in X$, $\beta_t(x)$ is the ordinary 1-dimensional Brownian motion.

We express probability and conditional probability, expectation and conditional expectation using P's and E's respectively. We denote the Fourier transform of functions and measures by the symbol $^\wedge$, convolution by $*$. We denote by K(X) the space of continuous functions on X with compact support with its canonical topology. We denote by E' the topological dual of a topological space E, by L(E,F) the space of continuous linear mappings of E into F. We denote by D the Schwartz space of test functions with D' the space of distributions. We use the indices, s,τ,c,e for spaces L(E,F) to provide them with the topologies of pointwise convergence, the Mackey topology, the topology of uniform convergence on compacts and the topology of uniform convergence on equicontinuous subsets, respectively. Thus $K'_s(X)$ is M(X), the space of random measures with the vague topology, i.e. a sequence μ_i converges vaguely to μ in M, written $\mu_i \xrightarrow{v} \mu$, if $\mu_i(\phi) \longrightarrow \mu(\phi)$ for all $\phi \in K$.

Let E and F denote locally convex topological vector spaces. We identify the algebraic tensor product $E' \otimes F$ with the space of finite rank continuous linear mappings of E into F. Assuming E and F to be complete, the tensor product $E \hat{\otimes}_e F$ is the completion, in the topology of $L_e(E'_\tau, F)$, of the finite rank weakly continuous mappings of E' into F ([5]).

We assume that there is no population dispersion, so that $A = 0$ in equation (1). All processes are adapted to probability spaces (Ω, F_t, μ), F_t an increasing system of sub-σ-fields of F. For convenience we assume that all random variables are square integrable. The results of the paper hold if $L^2(\Omega)$ is replaced by any semi-reflexive locally convex topological vector space of random variables.

Random distributions are distributions with values which are random variables. We identify them with linear mappings of test functions to random variables. We thus consider $L(D(X), L^2(\Omega))$. The natural topology for random distributions is that as elements of $D' \hat{\otimes}_e L^2$. Since D' has the approximation property, $D' \otimes L^2$ is dense in $L_c(D, L^2)$, so

$$D' \hat{\otimes}_e L^2 = L_e(D'_c, L^2) \simeq L_c(D', L^2) = \overline{D' \otimes L^2} \quad \text{(by [5] Proposition 35).}$$

Brownian motion in space-time is a random distribution $\beta_t(\phi)$ where $\phi(x)$ may be chosen in $K(X)$. The covariance kernel is in $D'(X \times X)$. Differentiating with respect to time we obtain space-time white noise, uncorrelated in space, i.e. $W_t(x) = \frac{\partial \beta_t}{\partial t}(x)$ formally represents a Gaussian generalised random process with zero mean and covariance kernel $\delta_{x-y} \in D'(X \times X)$. The following proposition is well known; the proof given here is coordinate-free.

PROPOSITION 1.

The white noise $W_t(x)$ is not in $L^2(X)$, where X has been provided with Lebesque measure.

<u>PROOF</u>

We know that if a square integrable random variable takes values in $L^2(X)$ then $E(\xi) \in L^2(X)$ and $E(\xi \otimes \xi) \in L^2(X \times X)$. If it were true that $W_t(x) \in L^2(X)$ its covariance kernel would be in L^1; however the covariance kernel δ_{x-y} is clearly not in $L^1(X \times X)$.

Since neither σ nor α are necessarily linear the equation (1) does not make sense even as an equation in random distributions. Instead of the equation (1) we consider a field of transition probabilities over the region X; this is a much more natural concept for population processes seeing that these are classically described by equations for their transition probabilities. Heuristically, the equation (1) is a continuous tensor product, over index set X, of ordinary stochastic differential equations (cf. [5] § 8).

2. Random measures

By Proposition 1 we are not able to find solutions to (1) using stochastic processes with state space $L^2(X)$. The approach in [1] was to use stochastic processes with state space $M(X)$. We shall instead consider random variable valued measures; these may be thought of as σ-additive random-variable valued functions on the Borel sets of X, or alternately as linear random variable valued functions on $M(X)$. Since population densities must be non-negative and non-negative distributions are measures, we are justified in considering those random distributions which are random measures and so to replace $D'(X)$ by $M(X)$. Since we shall want to approximate random measures by atomic random measures we need the analogue, for vector valued measures, of the vague topology for measures, i.e. we want the topology $L_s(K,L^2)$.

DEFINITION

Denote by $M \bar{\otimes}_s L^2$ the closure of $M \otimes L^2$ in $L_s(K,L^2)$. A continuous linear mapping n of K into L^2 is thus an element of $M \bar{\otimes}_s L^2$ if, for all $\varepsilon>0$ there exists a continuous linear mapping m of finite rank from K into L^2 such that, for all ϕ in K, $||n(\phi) - m(\phi)||_2 \leq \varepsilon$. We call the topology of $M \bar{\otimes}_s L^2$ the vague topology for random measures and use the same notation for convergence of sequences as for M.

PROPOSITION 2

The spaces $M \bar{\otimes}_s L^2$ and $M \hat{\otimes}_e L^2$ are incompatible.

<u>PROOF</u>

Since $M = K'_s$ its topology is that of $\prod_{\phi\in K} \mathbb{C}_\phi$, $\mathbb{C}_\phi = \mathbb{C}$. Given any $\varepsilon>0$, for any $\phi\in K$ one has $|\mu(\phi) - \mu_0(\phi)| < \varepsilon$ whenever μ is the neighbourhood $\{\mu : |\mu(\phi) - \mu_0(\phi)| < \varepsilon$ for all $\phi\in K\}$ of μ_0 in M. Thus K itself is an equicontinuous subset of M', so $L_s(K,L^2)$ and $L_e(K,L^2)$ are not homeomorphic.

We shall also be considering the narrow topology for the subspace of M^1 of probability measures in M viz. $m_k \Rightarrow m$ if $m_k(\phi) \to m(\phi)$ for all bounded continuous ϕ. (This is called the weak topology in some of the literature.) With respect to the vague topology M^1 is a Hausdorff space and a sequence in M^1 which converges vaguely to a limit remaining in M^1 also converges narrowly to this limit [7].

PROPOSITION 3

A σ-additive set function on the Borel sets of X taking values in $L^2(\Omega)$ and of finite variation $|n|$ defines an element of $M \bar{\otimes}_s L^2$ by means of the relation $n(\phi) = \int_X \phi(x)\, n(dx)$. Conversely an element of $M \bar{\otimes}_s L^2$ defines an L^2-valued measure. The usual definitions of atomicity are consistent for both viewpoints.

PROOF

We can approximate n in the vague topology by elements of $X \otimes L^2$. Indeed $L^2(\Omega)$ is reflexive so that the Random-Nikodym property holds and

$n(\phi) = \int_X \phi(x)\lambda(x)|n|(dx)$ where λ is a Bochner-integrable function; we approximate $\phi(x)\lambda(x)$ uniformly in $L^2(\Omega)$ by functions taking a finite number of values and approximate $|n|$ in M by atomic measures. Conversely, given an $n \in M \bar{\otimes}_s L^2$ we can define n(A) for a Borel set A in X. Indeed, for $f(w) \in L^2(\Omega)$, the inner product $(n(\phi)|f)$ belongs to M(X); hence we can define $(n(A)|f)$ and also n(A).

PROPOSITION 4

The finite rank atomic random measures are dense in $M \bar{\otimes}_s L^2$.

PROOF

Let $\psi \epsilon K$, $\mu_i \epsilon M$, $\phi_i \epsilon L^2$ for $i=1,\ldots,n$. Given $\varepsilon>0$ we can find atomic measures ν_i such that $\|\sum_1^n \mu_i(\psi)\ \phi_i - \sum_1^n \nu_i(\psi)\phi_i\|_2 < \varepsilon$. The proposition follows since $M \otimes L^2$ is everywhere dense in $M \bar{\otimes}_s L^2$.

REMARK

A random measure defines a characteristic functional on K, i.e. a positive definite linear functional Φ on K, with $\Phi(0) = 1$ and continuous on all finite dimensional subspaces, by means of the joint probability distributions for finite sets of functions in K. Denote by $\mathcal{M}$ the σ-ring generated by the cylinder sets in M. Using a theorem of Prohorov ([2] Theorem 1) a random measure n can be shown to define a probability measure on $(M,\mathcal{M})$; for a Borel set A in $\mathbb{R}^p$ the measure of the cylinder set

$$\{\mu : (\mu(\phi_1),\ldots,\mu(\phi_p)) \in A\}$$

is determined, via the joint probability distribution, as

$$P\{(n(\phi_1),\ldots,n(\phi_p)) \in A\}$$

3. Solutions for atomic initial conditions

Consider equation (1) with A = 0. We shall assume that n is a random measure process, i.e. each n_t is a random-variable valued measure, conditioned in that n_0 is fixed. By the remark in § 2 one may also consider n_t to be a measure valued process. We assume initial conditions $n_0 \epsilon M \bar{\otimes}_s L^2$ given. When n_0 is atomic with support S the solution n_t is obviously a set of diffusion processes $(n_t(x))_{x \epsilon S}$ such that the $n_t(x)$ are mutually independent and each satisfies the ordinary stochastic differential equation

$$\frac{d}{dt} n_t(x) = \alpha_x(n_t(x)) + \sigma_x(n_t(x)) \frac{d\beta_t(x)}{dt} \tag{2}$$

with $n_0(x)$ given. We have seen in §1 that for fixed x the process $\beta_t(x)$ is ordinary 1-dimensional Brownian motion. We interpret these equations in the Ito sense so that for every $x \in S$ the process $n_t(x)$ is a stationary Markov process with transition probability measure $\mu^t_{n_0(x)}$ i.e. if A is a Borel set in $\mathbb{R}^+$, then $\mu^t_{n_0(x)}(A)$ is the conditional probability $P\{n_t(x) \in A : n_0(x)\}$. It is determined from the Fokker-Planck equation

$$\frac{\partial p}{\partial t} = -\frac{\partial}{\partial y}(\alpha_x(y)p) + \frac{1}{2}\frac{\partial^2}{\partial y^2}(\sigma_x^2(y)p) \tag{3}$$

for its density functions $p(y,t)$. We shall see in Theorem 1 that for diffuse n_0 these transition probability measures are all degenerate. It follows from the CKS (Chapman-Kolmogorov-Smoluchovski) equations

$$\mu^{t+\tau}_{n_0(x)} = \mu^t_{n_\tau(x)} * \mu^\tau_{n_0(x)}$$

that for fixed x the $\mu^t_{n_0(x)}$ are mutually equivalent measures for all t. We denote the Fourier transform $\hat{\mu}^t_{n_0}$ by $\Phi^t_{n_0}$.

PROPOSITION 5

The Fourier transform of a transition probability measure is a conditional characteristic functional, i.e. $\Phi^t_{n_0(x)}(s)$ is the conditional expection $E_{n_0}\{e^{in_t(x)s}\}$.

PROOF

To see that the conditional expectation may indeed be defined in this way see [3]V §§10,11.

PROPOSITION 6

Assume, besides the usual conditions necessary for the formation of the Fokker-Planck equation, that the transition probability density function p for equation (1), and also $\frac{\partial p}{\partial y}$, vanish as y tends to infinity. Then the conditional characteristic functional $\Phi^t_{n_0}$ is the solution of the equation

$$\frac{\partial \phi}{\partial t} = -s\hat{\alpha}_x * \frac{\partial \phi}{\partial s} - s^2 \hat{\sigma}_x^2 * \phi \tag{4}$$

with initial condition $\phi(0) = \exp\{isn_0(x)\}$.

PROOF

We obtain equation (4) formally by taking the Fourier transform of the Fokker-Planck equation. To verify the initial condition we use Ito's formula [3] and see that

$$\Phi^t_{n_0}(s) = e^{in_0 s} - \tfrac{1}{2}s^2\int_0^t E_{n_0}(e^{in_t s}\sigma^2(n_t))dt + is\int_0^t E_{n_0}(e^{in_t s}\alpha(n_t))dt$$

The result follows putting t = 0.

We shall call equation (4) the CCF equation.

REMARK

It is reasonable to expect the condition $p \to 0$ as $y \to \infty$ if the population does not explode. Using the martingale inequality it can easily be seen that almost surely $p \to 0$ as $y \to \infty$ whenever α and σ are bounded.

We interpret equation (1) with atomic n_0 as the tensor product of the stochastic differential equations (2) with solution $n_t = \sum_{x\in S} n_t(x)$. For a set B in X, $n_t(B) = \sum_{x\in S\cap B} n_t(x)$, a sum of mutually independent random variables. For disjoint sets B in X the $n_t(B)$ are independent random processes. The transition probability for $n_t(B)$ is $\mathop{*}_{x\in S\cap B} \mu^t_{n_0(x)}$ and the characteristic function $\Phi^t_{n_0(B)} = \prod_{x\in S\cap B} \Phi^t_{n_0(x)}$. To see that n_t is also a measure valued process think of it as $\sum_{x\in S} n_t(x)\delta_x$. The transition probability of this process is $\bigotimes_{x\in S} \mu^t_{n_0(x)}$. To consider n_t as an atomic random measure define $n_t(\phi)$, where $\phi\in K$, to be $\sum_{x\in S} n_t(x)\phi(x)$. The conditional characteristic functional, $E_{n_0}\{e^{in_t(\phi)}\}$ or $\prod_{x\in S} \Phi^t_{n_0(x)}(\phi(x))$, shall be denoted by $\Phi^t_{n_0}(\phi)$.

4. Solutions for diffuse initial conditions

Let $n_0 \in M \bar{\otimes}_s L^2$ be a diffuse random measure and suppose that $n^{(j)}$ is a sequence of atomic random measures, with supports (X_j), which converges vaguely to n_0. Denote the solution of (1) for initial conditions $n^{(j)}$ by $n^{(j)}_t$. Denote the convolution product of the $\mu^t_{n^{(j)}(x)}$, $x\in X_j$, by $\mathop{*}_{x\in X_j} \mu^t_{n^{(j)}(x)}$. As in [1] we say that (1) with initial conditions n_0 is solvable if the transition probabilities of the

$n_t^{(j)}$, i.e. the $\bigotimes_{x\in X_j} \mu^t_{n^{(j)}(x)}$, converge narrowly. We wish to find random measure processes which satisfy equation (1); we call these (vague) solutions.

THEOREM

In order that equation (1), with A = 0 and n_0 a diffuse random measure in $M \bar{\otimes}_s L^2$, have a unique solution n_t it suffices, as $n^{(j)} \xrightarrow{v} n_0$, that for each Borel set B in X the $\Phi^t_{n^{(j)}(B)}$ converge uniformly on compacts, or that the sequence $\ast_{x\in X} \mu^t_{n^{(j)}(x)}$ converges narrowly in $M^1(R)$ to a limit, $\mu^t_{n_0}$ say. Then for each $x\in X$ necessarily $\mu^t_{n^{(j)}(x)} \Rightarrow \delta_0$. The random variable $n_t(B)$, the limit in probability of the approximating solutions, will be infinitely divisible. The solution is not necessarily a Markov process; it is weakly Markov in that $\mu^t_{n_0}$ satisfies the CKS equation only on sets with $\mu^t_{n_0}$ - null boundary.

PROOF

It follows directly from [7] Appendix, Theorem 6, that the conditions given are sufficient for the existence of the conditional characteristic functional $\Phi^t_{n_0(B)}$ and hence the random process $n_t(B)$. Since M^1 with the narrow topology is Hausdorff the solution will be unique. It is infinitely divisible since for disjoint sets B the $n_t(B)$ are independent random variables. The convergence of $\prod_{x\in X_j} (\Phi^t_{n^{(j)}(x)}(s))$ implies that as $j \to +\infty$ each $\Phi^t_{n^{(j)}(x)}(s) \to 1$ and so $\mu^t_{n^{(j)}(x)} \Rightarrow \delta_0$ (see [7] Appendix, Remark 2). A narrow limit of transition probabilities satisfying the CKS equation need not satisfy the CKS equation, so n_t is not necessarily Markov; since $\mu^{t+\tau}_{n^{(j)}} = \mu^t_{n^{(j)}} * \mu^\tau_{n^{(j)}}$ and, convolution being commutative, also $\ast_{x\in X_j} (\mu^{t+\tau}_{n^{(j)}(x)}) = (\ast_{x\in X_j} \mu^t_{n_t^{(j)}(x)}) * (\ast_{x\in X_j} \mu^\tau_{n^{(j)}(x)})$, so $\mu^{t+\tau}_{n_0}(f) = (\mu^t_{n_t} * \mu^\tau_{n_0})(f)$ for all bounded continuous f. The weak Markov property in the statement of the theorem follows since, for probability measures ν_i, ν on a locally compact Hausdorff space, $\nu_i \Rightarrow \nu$ if and only if $\nu_i(B) \to \nu(B)$ for all sets B with ν-null boundary.

Known results on infinitely divisible random distributions (cf.[5] Appendix E) imply the following:

COROLLARY 2

Under the conditions of the theorem above $n_t(B)$ is an infinitely divisible random variable, so $\Phi^t_{n_0(B)}$ is of the form $\exp\{\int_X F^t_x(s)d\nu(x)\}$ for some measure ν on X and conditionally positive definite function F^t. The conditional characteristic functional for the random measure n_t, i.e. $\Phi^t_{n_0}$ where $\Phi^t_{n_0}(\phi) = E_{n_0}(e^{in_t(\phi)})$, is given by $\Phi^t_{n_0}(\phi) = \exp\int_X F^t_x(\phi(x))d\nu(x)$.

To verify solvability in [1], X compact, it was sufficient, by a theorem of Prohorov, to verify that the $\Phi^t_{n^{(j)}}$ converge pointwise in K(X). The limit of the $\phi^t_{n^{(j)}}$ is not necessarily continuous on K.

If n_0 is an arbitrary random measure in $M \bar{\otimes}_s L^2$ and if α and σ are linear we may partition n_0 into atomic and diffuse parts and superimpose the solutions for these parts.

<u>REMARK</u>

The solutions of equation (1) are what we call vague solutions. Indeed $\Phi^t_{n_0}$ does not in general determine a random measure, even if it were continuous. From another viewpoint, $n_t(B)$ does not in general determine $n_t(\phi)$ since it is not necessarily in any topological vector space with the Radon-Nikodym property and Proposition 3 does not apply.

5. <u>Examples</u>; randomly disturbed Malthusian equations

(a) Brownian motion; $\frac{\partial n}{\partial t} = \frac{\partial \beta}{\partial t}$.

The Fokker-Planck equation is $\frac{\partial p}{\partial t} = +\frac{1}{2}\frac{\partial^2 p}{\partial y^2}$ and the CCF equation is $\frac{\partial \Phi}{\partial t} = -\frac{1}{2} s^2\Phi$.

Thus $\Phi^t_{n_0(x)}(s) = \exp\{in_0(x)s - s^2t\}$. By Corollary 2 it follows that

$\phi^t_{n_0}(\phi) = \exp\{in_0(\phi) - t\int_X |\phi(x)|^2 d\nu(x)\}$ for some measure ν on X determined by n_0.

(b) The Langevin equation $\frac{\partial n}{\partial t} = \alpha n + \sigma\frac{\partial \beta}{\partial t}$, α, $\sigma \in \mathbb{R}$, $\alpha \neq 0$.

The Fokker-Planck equation is

$$\frac{\partial p}{\partial t} = -\alpha \frac{\partial}{\partial y}(xp) + \frac{1}{2}\sigma\frac{\partial^2 p}{\partial y^2}$$

with CCF equation

$$\frac{\partial\Phi}{\partial t} + s\,\frac{\partial\Phi}{\partial s} + \tfrac{1}{2}\,\sigma\,s^2\,\phi = 0.$$

Solving the related equations

$$dt = -ds/\alpha s = -2d\Phi/s^2\Phi$$

we obtain

$$\Phi^t_{n_0}(s) = \exp\{isn_0 e^{\alpha t} - \tfrac{1}{4}\,\sigma s^2\alpha^{-1}(e^{2\alpha t}-1)\}$$

Thus

$$\Phi^t_n(\phi) = \exp\{ie^{\alpha t}n_0(\phi) - \tfrac{1}{4}\,\sigma\alpha^{-1}(e^{2\alpha t}-1)\int_X |\phi(x)|^2 d\nu(x)\}$$

(c) The continuous branching process $\frac{\partial n}{\partial t} = \alpha n_t + \sqrt{2\gamma n}\,\frac{\partial\beta}{\partial t}$, where $\alpha\in\mathbb{R}$, $\gamma > 0$.

As in [1], $\Phi^t_{n_0(x)}(s) = \exp\{i\alpha s n_0(x)e^{\alpha t}[\alpha - i\gamma s(e^{\alpha t}-1)]^{-1}\}$,

$$\Phi^t_{n_0}(\phi) = \exp\{i\int \alpha\phi(x)e^{\alpha t}[\alpha - i\gamma\phi(x)(e^{\alpha t}-1)]^{-1}n_0(dx)\}.$$

(d) $\frac{\partial n}{\partial t} = \alpha n + \sigma n\,\frac{\partial\beta}{\partial t}$ where $\alpha,\sigma \in \mathbb{R}$, $\sigma \neq 0$.

We assume X compact and that the domain of n_0 can be extended from $K(X)$ to $L^2(X)$. The Fokker-Planck and CCF equations are respectively

$$\frac{\partial p}{\partial t} = -\alpha\,\frac{\partial}{\partial y}\,(yp) + \tfrac{1}{2}\,\sigma^2\,\frac{\partial^2}{\partial y^2}\,(y^2 p)$$

$$\frac{\partial\Phi}{\partial t} = -\alpha\,s\,\frac{\partial\Phi}{\partial s} - \tfrac{1}{2}\,\sigma^2 s^2\,\frac{\partial^2\Phi}{\partial s^2}.$$

Substituting $|s| = e^\theta$ and $\Phi = e^{-k\theta}\Psi$ where $k = 1-2\alpha\sigma^{-2}$ we obtain the equation

$$\sigma^2\,\frac{\partial\Psi}{\partial\theta^2} - h\Psi = \frac{\partial\Psi}{\partial t}$$

where $h = h(\alpha,\sigma)$. Its fundamental solution is

$$(4\pi\sigma^2 t)^{-\frac{1}{2}}\exp\{-ht-(\xi-\theta)^2/4\sigma^2 t\}.$$

Applying the initial condition we obtain

$$\Phi^t_{n_0(x)}(\theta) = e^{-ht-k\theta}\int_{\mathbb{R}} e^{in_0(x)\xi}\,d\nu^t_\theta(\xi)$$

where $d\nu^t_\theta$ is a gaussian measure on $\mathbb{R}$. Thus $\Phi^t_{n_0}(\phi)$ is of the form

$$e^{-ht}\,\tilde{\prod_{x\in X}}\,|\phi(x)|^{-k}\int_{L^2(X)} e^{in_0(\psi)}\,d\nu^t_\phi(\psi)$$

where $\tilde{\Pi}$ denotes a continuous product (see [6] §8.1) and ν_ϕ^t denotes a Gaussian measure on $L^2(X)$.

Consider the case $\alpha=0$. As in [1] Theorem 6.2, the conditional variance of $n_t(x)$ is $(n_0(x))^2(e^t-1)$. Let B be a Borel set in X. Assuming that $n_0(x) = j^{-1}n_0(B)$ for each $x \in X_j \cap B$, the conditional variance

$$E_{n^{(j)}}[n_t^{(j)}(B) - E_{n^{(j)}}n_t^{(j)}(B)]^2$$

$$= \sum_{x\in X_j\cap B} E_{n^{(j)}}[n_t^{(j)}(x) - E_{n^{(j)}}n_t^{(j)}(x)]^2$$

$$= j(j^{-1}n_0(B))^2(e^t-1) \longrightarrow 0 \text{ as } j \longrightarrow \infty.$$

Thus n_t is stationary, i.e. for $t\geq 0$ almost surely $n_t \equiv n_0$.

BIBLIOGRAPHY

1. Dawson D.A. (1975) "Stochastic evolution equations and related measure processes". J. Mult. Anal. 5 1-52.

2. Cartier P. (1963/4) Processus aleatoires generalisés. Seminaire Bourbaki.

3. Feller W. (1971) An Introduction to probability theory and its applications Vol. II Second edition. Wiley, New York.

4. Gihman I.I. and Skorohod A.V. (1972) Stochastic differential equations, Springer-Verlag, Berlin.

5. Grothendieck A. (1965) Produits tensoriels et espaces topologiques nucleaires, Mem. Amer. Math. Soc. 16.

6. Guichardet A. (1972) Symmetric Hilbert spaces and related topics. Lecture notes in mathematics, 261. Springer-Verlag.

7. Schwartz L. (1975) Radon measures. Studies in Maths. No. 6 Tata Inst. Fund. Research Bombay.

STRONG MEASURABILITY, LIFTINGS AND THE CHOQUET-EDGAR THEOREM

Heinrich von Weizsäcker

The idea of liftings of abstract valued functions with relatively compact range (developed by A. and C. Ionescu Tulcea in |6|,p. 50ff) is applied in the first part of this note to the regularization of functions with Radon image measures and arbitrary (completely regular) range. These functions generalize the strongly measurable Banach space valued functions. The regularization defines in a natural way a 'lifting' from the weak equivalence classes to the strong equivalence classes of these functions (see Theorem 1.4 and Cor. 1.6 below).

In the second part we sketch how this regularization or similar techniques can be used in the proof of results concerning maximal integral representations in bounded sets in locally convex spaces. These results cover both the compact situation of the classical Choquet theory and the results of Edgar |3| on sets with the Radon-Nikodym-Property in Banach spaces. A detailed study of a slightly more general version of Theorem 2.2 below has been given in |10|, §III.

Part I : Strong measurability and liftings.

There are several alternatives how to generalize the concept of strong measurability of Banach space valued functions to the case of general image spaces. Here are two (cf. 1.2 and 1.3 below):

1.1 Notation. Let (Ω,Σ,P) be a complete probability space and let E be a completely regular Hausdorff space with its Borel σ-algebra $\mathcal{B}(E)$ and its Baire σ-algebra $\mathcal{B}_o(E)$. Let $\mathcal{P}_t(E)$ be the set of all Radon probability measures (i.e. compact regular) on $\mathcal{B}(E)$. $C_b(E)$ denotes the Banach space of real bounded continuous functions on E.

1.2 Definition. $L^o(\Omega,\Sigma,P,E)$ denotes the set of all Σ- $\mathcal{B}(E)$ measurable functions $\phi : \Omega \longrightarrow E$ whose image measure belongs to $\mathcal{P}_t(E)$. This image measure will be denoted by ϕP.

1.3 Definition. $L^o_o(\Omega,\Sigma,P,E)$ denotes the set of all Σ - $\mathcal{B}_o(E)$ measurable functions $\phi : \Omega \longrightarrow E$ whose image maesure on $\mathcal{B}_o(E)$ can be extended to an element of $\mathcal{P}_t(E)$.

There are good reasons to prefer the second definition. One is the fact that for a topological group $(E,+)$ the space $L^0_o(\Omega,\Sigma,P,E)$ is closed under pointwise addition whereas in general $L^0(\Omega,\Sigma,P,E)$ is not (J. Pachl, unpublished). However we shall see in the following theorem that one can associate in a smooth way to any element of $L^0_o(\Omega,\Sigma,P,E)$ an element of $L^0(\Omega,\Sigma,P,E)$ and the stronger measurability properties of this new function will simplify some arguments in the sequel.

1.4 Theorem. Let (Ω,Σ,P) be a complete probability space and let E be a completely regular Hausdorff space. Let $\rho: \mathcal{L}^\infty(\Omega,\Sigma,P) \longrightarrow \mathcal{L}^\infty(\Omega,\Sigma,P)$ be a multiplicative lifting. Then

a) For any $\phi \in L^0_o(\Omega,\Sigma,P,E)$ there is a ('weakly equivalent') function ϕ^ρ in $L^0(\Omega,\Sigma,P,E)$ satisfying

$$(1) \qquad P\Big(\bigcup_{f \in C_b(E)} \{ \omega : f(\phi^\rho(\omega)) \neq \rho(f\circ\phi)(\omega) \} \Big) = 0.$$

ϕ^ρ is uniquely determined outside the nullset in (1).

b) The functions ϕ^ρ in a) have the following properties:

b1) If $h : E \longrightarrow F$ is continuous where F is completely regular Hausdorff, then

$$(2) \qquad (h\circ\phi)^\rho = h \circ \phi^\rho \qquad P\text{- a.e.}$$

holds for all $\phi \in L^0_o(\Omega,\Sigma,P,E)$. The same is true if h is ϕP - Lusin measurable and $\phi \in L^0(\Omega,\Sigma,P,E)$.

b2) Let $(E_k)_{k=1,2,..}$ be a sequence of completely regular Hausdorff spaces. If ϕ_k is in $L^0_o(\Omega,\Sigma,P,E_k)$ for all k then the 'product map' $<\phi_1,\phi_2,...>$ is in the set $L^0_o(\Omega,\Sigma,P, \prod_{k\in\mathbb{N}} E_k)$ and we have

$$(3) \qquad (<\phi_1,\phi_2,....>)^\rho = <\phi_1^\rho,\phi_2^\rho,....> \qquad P\text{- a.e. .}$$

In particular the map $<\phi_1^\rho,\phi_2^\rho,....>$ is even Σ - $\mathcal{B}(\prod_{k\in\mathbb{N}} E_k)$ measurable.

Proof. a) Suppose first E to be compact. For $\phi \in L^0_o(\Omega,\Sigma,P,E)$ define a map $\rho'(\phi)$ from Ω to E by

$$(4) \qquad f(\rho'(\phi)(\omega)) = \rho(f\circ\phi)(\omega) \qquad (\omega \in\Omega,\ f \in C_b(E)).$$

$\rho'(\phi)$ is well defined and Σ - $\mathcal{B}(E)$ measurable by (|6|,p.51 Theorem 1 and p.52 remark c). The argument used there for the Σ-$\mathcal{B}(E)$ measurability shows also that the image measure of $\rho'(\phi)$ is τ-smooth and hence in $\mathcal{P}_t(E)$ since E is compact. So $\rho'(\phi)$ is in $L^0(\Omega,\Sigma,P,E)$ and satisfies (1).

Now let E be any completely regular Hausdorff space and let $\phi \in L^0_o(\Omega,\Sigma,P,E)$ be given. We consider E as a subset of its Stone-Čech compactification βE. Then ϕ is also in $L^0_o(\Omega,\Sigma,P,\beta E)$. The function $\rho'(\phi)$ defined by (3) is in $L^0(\Omega,\Sigma,P,\beta E)$ and its image measure $\rho'(\phi)P$ is the Radon measure extension of the Baire image measure of ϕ. From this and the definition of $L_o^o(\Omega,\Sigma,P,E)$ we conclude that there is a σ-compact subset K of E such that $P\{\omega : \rho'(\phi)(\omega) \in K\} = \rho'(\phi)P(K) = 1$. Thus, any function $\phi^\rho : \Omega \to E$ such that $\phi^\rho(\omega) = \rho'(\phi)(\omega)$ if $\rho'(\phi)(\omega) \in K$, satisfies (1) and is in $L^o(\Omega,\Sigma,P,E)$ because of (4) and $\rho'(\phi) \in L^o(\Omega,\Sigma,P,\beta E)$. The uniqueness statement in a) follows from the fact that $C_b(E)$ separates points in E.

b) 1. Let F, $h : E \to F$ continuous, $\phi \in L_o^o(\Omega,\Sigma,P,E)$, ϕ^ρ and $(ho\phi)^\rho$ be given. Then there is a P-nullset $N \in \Sigma$ such that

$$f(\phi^\rho(\omega)) = \rho(fo\phi)(\omega)$$

and

$$g((ho\phi)^\rho(\omega)) = \rho(goho\phi)(\omega)$$

hold for all $f \in C_b(E)$, $g \in C_b(F)$ and $\omega \in \Omega\setminus N$. Letting $f = goh$ we get

$$g((ho\phi^\rho)(\omega)) = \rho(goho\phi)(\omega) = g((ho\phi)^\rho(\omega))$$

for all $g \in C_b(F)$ and hence $ho\phi^\rho(\omega) = (ho\phi)^\rho(\omega)$ for all $\omega \notin N$.

Now assume ϕ to be in $L^o(\Omega,\Sigma,P,E)$ and h to be ϕP- Lusin measurable. Then there is a sequence (K_n) of disjoint compact subsets of E satisfying $\phi P(\bigcup_{n\in\mathbb{N}} K_n) = 1$ such that the restrictions $h|K_n$ are continuous for all n. Denote by η the original topology on E and by τ the sum topology induced by therepresentation

$$E = \bigcup_{n\in\mathbb{N}} K_n \cup \bigsqcup_{x\notin\bigcup K_n} \{x\} .$$

Then ϕ is also in $L^o(\Omega,\Sigma,P,(E,\tau))$ since (Ω,Σ,P) is complete. Applying the first part of this proof to the continuous map $id : (E,\tau) \to (E,\eta)$ we conclude that the function ϕ^ρ changes only on a P-nullset if we interchange the topologies η and τ. Since $h : (E,\tau) \to F$ is continuous, we get (2) also in this situation.

2. For the proof of (3) let the sequences (E_k) and (ϕ_k) be given as indicated. The lemma formulated below implies that the map $<\phi_1,\phi_2,\ldots>$ is in $L_o^o(\Omega,\Sigma,P,\prod_{k\in\mathbb{N}} E_k)$. Let N be the countable union of the P-nullsets in (1) corresponding to the functions $(<\phi_1,\phi_2,\ldots>)^\rho$ and ϕ_k^ρ $(k\in\mathbb{N})$. If $pr_n : \prod_{k\in\mathbb{N}} E_k \to E_n$ denotes the projection we get

$$\begin{aligned} fopr_n o(<\phi_1,\phi_2,\ldots>)^\rho(\omega) &= \rho(fopr_n o<\phi_1,\phi_2,\ldots>)(\omega) \\ &= \rho(fo\phi_n)(\omega) = fo\phi_n^\rho(\omega) = fopr_n o<\phi_1^\rho,\phi_2^\rho,\ldots>(\omega) \end{aligned}$$

for all $\omega \in \Omega\setminus N$, $n \in \mathbb{N}$ and $f \in C_b(E_n)$. Now the functions $fopr_n$ $(n \in \mathbb{N}, f \in C_b(E_n))$ separate the points of $\prod E_k$, so the proof of (3) and hence of the theorem is complete.

In the last part of the proof we used the follwing lemma

<u>1.5 Lemma</u>. Let $(E_k)_{k=1,2,..}$ be a sequence of completely regular Hausdorff

spaces. Let μ be a probability measure on the product σ-algebra $\bigotimes_{k\in\mathbb{N}} \mathcal{B}_o(E_k)$ whose marginal measures on the σ-algebras $\mathcal{B}_o(E_k)$ have Radon measure extensions. Then $\mathcal{B}_o(\prod_{k\in\mathbb{N}} E_k)$ is contained in the μ-completion of $\bigotimes_{k\in\mathbb{N}} \mathcal{B}_o(E_k)$ and μ has a Radon measure extension.

Proof: It suffices to prove that every $f \in C_b(\prod E_k)$ is μ-measurable, the rest follows from standard arguments. For $\varepsilon > 0, k \in \mathbb{N}$ choose $K_k \subset E_k$ compact such that $\mu_k(K_k) > 1 - \varepsilon 2^{-k}$ where μ_k denotes the Radon extension of the k-th marginal measure of μ. Then $K = K_1 \times K_2 \times \ldots$ is compact. By Stone-Weierstraß there is a $\bigotimes_{k\in\mathbb{N}} \mathcal{B}_o(E_k)$ - measurable continuous function f^ε such that $|f - f^\varepsilon| < \varepsilon$ on K. There are a number $m \in \mathbb{N}$ and open neighbourhoods $U_1,\ldots,U_m$ of $K_1,\ldots,K_m$ respectively such that

$$K \subset U^\varepsilon := U_1 \times \ldots \times U_m \times \prod_{k>m} E_k \subset \{x : |f(x) - f^\varepsilon(x)| < \varepsilon\}.$$

Each U_i may be chosen $\mathcal{B}_o(E_i)$- measurable since E_i is completely regular. This implies $U^\varepsilon \in \bigotimes \mathcal{B}_o(E_k)$. Also $\mu(U^\varepsilon) > 1 - \varepsilon$. For $\varepsilon_n = 2^{-n}$ and $U = \bigcap_{r\in\mathbb{N}} \bigcup_{n>r} U^{\varepsilon_n}$ we get $\mu(U) = 1$ and $f(x) = \lim_n f^{\varepsilon_n}(x)$ on U. Thus f is μ-measurable.

From now on E is a real locally convex Hausdorff space. We write E' (resp. $E^\times$) for its topological (resp. algebraic) dual space. For ϕ,ψ in $L_o^o(\Omega,\Sigma,P,E)$ we write $\phi \equiv \psi$ if $\langle y,\phi(.)\rangle = \langle y,\psi(.)\rangle$ P- a.e. for every $y \in E'$.

1.6 Corollary. Let E be a locally convex Hausdorff space. Then there is a map $T : L_o^o(\Omega,\Sigma,P,E) \to L^o(\Omega,\Sigma,P,E)$ with the properties

I.	$T\phi \equiv \phi$		$(\ \phi \in L_o^o(\Omega,\Sigma,P,E)\ )$
II.	$\phi \equiv \psi \Rightarrow T\phi = T\psi$	P- a.e.	$(\ \phi,\psi \in L_o^o(\Omega,\Sigma,P,E)\ )$
III.	$T(a\phi + b\psi) = aT\phi + bT\psi$	P- a.e.	$(\ \phi,\psi \in L_o^o(\Omega,\Sigma,P,E)$, a and b real).

Proof: For every ϕ choose $T\phi = \phi^\rho$ according to the theorem. Then III. holds by Theorem 1.4b). Outside the nullset in Theorem 1.4a) we get $f\circ T\phi = f\circ\phi^\rho = \rho(f\circ\phi) = f\circ\phi$ P- a.e. for all $f \in \{-n \vee (y \wedge n) : y \in E', n \in \mathbb{N}\}$, hence for all $f \in E'$. Similarly, if $\phi \equiv \psi$ then $y\circ\phi = y\circ\psi$ for all $y \in E'$ outside the two nullsets corresponding to ϕ and ψ. Since E' separates points this yields II.

Remark. Using an argument as in the proof of Corollary 1.7 below it is possible to substitute the space $L_o^o(\Omega,\Sigma,P,E)$ in 1.6 by the space of all scalarly measurable functions from Ω to E whose image measure on the σ-algebra generated by E' can be extended to a Radon measure on E.

In the sequel we shall need only the easier part a) of Theorem 1.4. Instead of this we could use the (deeper) results of |4|.

For finite $F \subset E'$ denote by $\sigma(F)$ the σ-algebra generated by F over E. A cylindrical measure on E may be considered as a set function ζ on the algebra

$$\underline{Z}(E) = \bigcup \{ \sigma(F) : F \subset E', F \text{ finite} \}$$

which is a probability measure on each $\sigma(F)$. We say that the cylindrical measure is concentrated on a set $C \subset E$ if there is a Radon measure $\mu \in \mathcal{P}_t(C)$ such that $\zeta(Z) = \mu(C \cap Z)$ for all $Z \in \underline{Z}$. A vector measure $m : \Sigma \to E'^{\times}$ with $m \ll P$ defines a cylindrical measure ζ_m by

$$\zeta_m \{x : (\langle y,x\rangle)_{y \in F} \in B\} = P\{\omega : (\frac{d\langle y,m(.)\rangle}{dP}(\omega))_{y\in F} \in B\}$$

where $F \subset E'$ is finite and B is a Borel set in $\mathbb{R}^F$.

1.7 Corollary: Let C be a subset of a locally convex Hausdorff space E. Let (Ω,Σ,P) be a complete probability space and let $m : \Sigma \to E'^{\times}$ be a vector measure $\ll P$. Then there is a density $dm/dP \in L^o(\Omega,\Sigma,P,C)$ iff the cylindrical measure ζ_m is concentrated on C.

Proof. If $\psi = dm/dP \in L^o(\Omega,\Sigma,P,C)$ exists then the image measure $\psi P \in \mathcal{P}_t(C)$ satisfies $\zeta_m(Z) = \psi P(C \cap Z)$ for all $Z \in \underline{Z}$, so ζ_m is concentrated on C.
For the proof of the converse let G be the compact product space $[-\infty,+\infty]^{E'}$. For each $y \in E'$ let ϕ_y be a P-density of the scalar measure $\langle y,m(.)\rangle$. By Stone - Weierstraß we have $\bigotimes_{y\in E'} \mathcal{B}([-\infty,+\infty]) = \mathcal{B}_o(G)$ and therefore $\phi = (\phi_y)_{y\in E'}$ is an element of $L^o_o(\Omega,\Sigma,P,G)$. By Theorem 1.4a) we may choose ϕ even in $L^o(\Omega,\Sigma,P,G)$. Our assumption on the cylindrical measure ζ_m gives us a measure $\mu \in \mathcal{P}_t(C)$ such that

$$(+) \qquad \mu\{x \in C : (\langle y,x\rangle)_{y\in F} \in B\} = P\{\omega : (\phi_y(\omega))_{y\in F} \in B\}$$

is true for all finite $F \subset E'$ and $B \in \mathcal{B}(\mathbb{R}^F)$ or even $B \in \mathcal{B}([-\infty,+\infty]^F)$. Let $j : E \to G$, $x \mapsto (\langle y,x\rangle)_{y\in E'}$ denote the canonical embedding. By (+) the Radon image measures ϕP and $j\mu$ are equal, since they coincide on all sets which depend only on finitely many coordinates. Thus, there is a σ-compact subset K of C satisfying $P\{\omega : \phi(\omega) \in jK\} = \phi P(jK) = j\mu(jK) = \mu(K) = 1$. On jK the inverse j^{-1} is Lusin measurable, hence $\psi(\omega) = j^{-1}(\phi(\omega))$ defines P- a.e. a P-density $\psi \in L^o(\Omega,\Sigma,P,C)$ of m.

Part II : On the Choquet-Edgar Theorem.

The following result has a long history (see e.g. |2| and the literature quoted there). It explains the intimate connection between martingale convergence and Choquet type theorems (see below).

2.1 Theorem. Let T be a completely regular Hausdorff space. Let $S \subset C_b(T)$ be a point separating cone such that $1 \in S$ and $\max(f,g) \in S$ whenever $f \in S, g \in S$. Then for any net $(\mu_i)_{i\in I}$ in $\mathcal{P}_t(T)$ the following are equivalent:

a) For every $f \in S$ the net $(\int_T f\, d\mu_i)_{i\in I}$ is increasing.

b) There are a probability space (Ω,Σ,P), an increasing net $(\Sigma_i)_{i\in I}$ of sub-σ-algebras of Σ and functions $\phi_i \in L^o(\Omega, \Sigma_i,P,T)$ $(i\in I)$ such that $\mu_i = \phi_i P$ for every $i\in I$ and $(f\circ\phi_i,\Sigma_i)_{i\in I}$ is a submartingale for every $f \in S$ (i.e. we have $\int_A f\circ\phi_i\, dP \leqq \int_A f\circ\phi_j\, dP$ whenever $A \in \Sigma_i$ and $i \leq j$).

Proof. The implication b) $\Rightarrow$ a) is obvious. The converse has been proved by Doob (|2|,Theorem 4.2) for T compact and the ϕ_i only in $L_o^o(\Omega,\Sigma_i,P,T)$. In the general situation let $\tilde{f}$ denote the canonical extension of a function $f \in C_b(T)$ to the Stone-Čech compactification βT of T and let $\tilde{\mu}_i$ be the Radon measure on βT induced by μ_i. Then Doob's result applied to the cone $\tilde{S} = \{ \tilde{f} : f \in S \}$ and the net $(\tilde{\mu}_i)$ yields a probability space (Ω,Σ,P), an increasing net $(\Sigma_i)_{i\in I}$ of sub-σ-algebras of Σ and functions $\tilde{\phi}_i \in L_o^o(\Omega,\Sigma_i,P,\beta T)$ $(i\in I)$ such that $\tilde{\mu}_i = \tilde{\phi}_i P$ for every $i\in I$ and $(\tilde{f}\circ\tilde{\phi}_i,\Sigma_i)_{i\in I}$ is a submartingale for every $\tilde{f} \in \tilde{S}$. We may assume (Ω,Σ_i,P) to be complete for every $i\in I$. Hence by Theorem 1.4a) it is possible to choose $\tilde{\phi}_i \in L^o(\Omega,\phi_i,P,\beta T)$. But then there is for each i a σ-compact set $K_i \subset T$ such that $P(\tilde{\phi}_i \in K_i) = \tilde{\mu}_i(K_i) = 1$. Therefore we have to change the $\tilde{\phi}_i$ only on P-nullsets in order to get a T-valued family $(\phi_i)_{i\in I}$ with the desired properties.

We shall apply this theorem in the case where T is a bounded (not necessarily convex) subset C of a locally convex Hausdorff space E and S is the cone of all f which can be extended to bounded continuous convex functions on the convex hull of C. In this situation the condition in Theorem 2.1b) is equivalent to the condition that $(\phi_i,\Sigma_i)_{i\in I}$ is a martingale. One implication follows from the fact that for every $y \in E'$ the restriction $y|E'$ is in $S \cap (-S)$. The converse results from Jensen's inequality which can be proved exactly as in the finite dimensional case using an Hahn-Banach separation argument.

The Choquet-ordering $\prec$ on $\mathcal{P}_t(C)$ is defined by: $\mu \prec \nu$ iff $\int f\, d\mu \leqq \int f\, d\nu$ for all $f \in S$. For the (straightforward) proof of antisymmetry cf. |5|,prop. 2.1 or |10|,p. 64. Theorem 2.1 and the preceding remark show that $\prec$ can also be described in terms of conditional expectations. Other descriptions using the concept of 'μ-dilations' are given in |5|, Theorem 2.2 and in |10|, p. 63.

For $\mu \in \mathcal{P}_t(C)$ the resultant $r(\mu)$ is defined by $\langle y,r(\mu)\rangle = \int_C \langle y,x\rangle d\mu(x)$ $(y \in E')$. We write $\mathcal{P}_{t,r}(C)$ for the set $\{ \mu \in \mathcal{P}_t(C) : r(\mu) \in C\}$. C is called measure convex (t-convex in |5|) if $\mathcal{P}_t(C) = \mathcal{P}_{t,r}(C)$. $\mathcal{P}_t(C)$ is endowed with the topology of pointwise convergence on $C_b(C)$.

2.2 Theorem. Let C be a bounded subset of a locally convex Hausdorff space E. Consider the following conditions:

① For every complete probability space (Ω,Σ,P) and every martingale $(\phi_i,\Sigma_i)_{i\in I}$ in $L^0(\Omega,\Sigma,P,C)$ there is a function $\phi \in L^0(\Omega,\Sigma,P,C)$ such that

$$\int_\Omega \gamma(\phi_i(\omega) - \phi(\omega))\, dP(\omega) \underset{i\in I}{\longrightarrow} 0$$

for every continuous seminorm γ on E.

② Every directed subset in $(\mathcal{P}_t(C),\prec)$ converges to its least upper bound.

③ Every increasing chain in $(\mathcal{P}_t(C),\prec)$ converges to its least upper bound.

④ C has the following Radon-Nikodym Property: Let (Ω,Σ,P) be a probability space with an algebra $\Delta\subset\Sigma$ such that Σ is the P-complete σ-algebra generated by Δ. Let $m : \Delta \to E$ be additive such that $\{m(D)/P(D) : D\in\Delta, P(D) > 0\} \subset C$. Then there is $\phi \in L^0(\Omega,\Sigma,P,C)$ satisfying $m(D) = \int_D \phi(\omega)\, dP(\omega)$ for all $D \in \Delta$.

Then ①, ②, ③ are equivalent and imply ④. If C is measure convex then all four conditions are equivalent.

Proof. ① ⇒ ③. Let $(\mu_i)_{i\in I}$ be an increasing chain in $(\mathcal{P}_t(C),\prec)$. By Theorem 2.1 there is a martingale (ϕ_i,Σ_i) on some probability space (Ω,Σ,P) such that $\mu_i = \phi_i P$ for all $i\in I$. We may assume that Σ is the P-completion of the σ-algebra generated by the union $\bigcup \Sigma_i$. Choose $\phi \in L^0(\Omega,\Sigma,P,C)$ according to ①. Let μ be the image measure ϕP. Then we get

$$\lim_I \int_C f\, d\mu_i = \lim_I \int_\Omega f\circ\phi_i\, dP = \int_\Omega f\circ\phi\, dP = \int_C f\, d\mu$$

first for all uniformly continuous $f \in C_b(C)$ and then for all $f \in C_b(C)$ by (|9|,p.40), in particular for all $f \in S$. This shows that μ is the least upper bound of (μ_i).

③ ⇒ ②. Let $(H,\leq)$ be a partially ordered directed set. It is easy to construct to any infinite subset G of H a directed set G' such that $G\subset G'\subset H$ and card G = card G'. Hence there is an increasing chain of directed subsets (G_i) of H satisfying $H = \bigcup_I G_i$ and card $G_i <$ card H for all i, if H is infinite. Using this observation it clear how to prove ② using ③ and transfinite induction on the cardinality of the directed set in ②.

② ⇒ ①. Let the martingale $(\phi_i,\Sigma_i)_{i\in I}$ on (Ω,Σ,P) be given. Again we may assume that Σ is the complete σ-algebra generated by $\bigcup\Sigma_i$. Then according to Theorem 2.1 the net $(\phi_i P)_{i\in I}$ is increasing in $(\mathcal{P}_t(C),\prec)$. Let μ be its limit and least upper bound. For each $y \in E'$ the martingale $(\langle y,\phi_i(.)\rangle)_I$ converges to a function $\phi_y \in \mathcal{L}^\infty(\Omega,\Sigma,P)$ in mean. Define a vector measure $m : \Sigma \to E'^{\times}$ by

$$\langle y,m(A)\rangle = \int_A \phi_y(\omega)\, dP(\omega).$$

For every finite Family $F = (y_1,\dots,y_n)$ in E' denote by pr_F the map $x \mapsto (\langle y,x\rangle)_F$. For each $f \in C_b(\mathbb{R}^F)$ we get

$$\begin{aligned}\int_C f o pr_F\, d\mu &= \lim_I \int_\Omega f o pr_F o \phi_i\,(\omega)\, dP(\omega)\\ &= \lim_I \int_\Omega f(\langle y_1,\phi_i(\omega)\rangle,\dots,\langle y_n,\phi_i(\omega)\rangle)\, dP(\omega)\\ &= \int_\Omega f(\phi_{y_1}(\omega),\dots,\phi_{y_n}(\omega))\, dP(\omega).\end{aligned}$$

Since μ is in $\mathcal{P}_t(C)$ it follows that the cylindrical measure defined by the vector measure m is concentrated on C. By corollary 1.7 there is a density $\phi = dm/dP \in L^o(\Omega,\Sigma,P)$. For each $i \in I$ we have

$$\int_A \langle y,\phi(\omega)\rangle\, dP(\omega) = \int_A \phi_y(\omega)\, dP(\omega) = \int_A \phi_i(\omega)\, dP(\omega) \qquad (A \in \Sigma_i),$$

so $\phi_i = E(\phi|\Sigma_i)$. Now let γ be a continuous seminorm on E and let π_γ be the projection into the Banach space defined by γ. Then $(\pi_\gamma o \phi_i,\Sigma_i)_{i\in I}$ is a uniformly bounded Banch space valued strongly measurable martingale with the 'closing' function $\pi_\gamma o \phi$. This implies

$$\int_\Omega \gamma(\phi_i(\omega) - \phi(\omega))\, dP(\omega) = \int_\Omega \| \pi_\gamma o \phi_i - \pi_\gamma o \phi \|_\gamma\, dP \xrightarrow[i \in I]{} 0$$

where $\|\,.\,\|_\gamma$ is the norm of that Banch space.

The proofs of ① ⇒ ④ and ④ ⇒ ① (if C is measure convex) are straightforward. For ① ⇒ ④ note that for a probability space (Ω,Σ,P) every finitely additive set function vanishing on the P-nullsets defines in the usual way a martingale with the finite measurable partitions as index set. For ④ ⇒ ① note that every martingale (ϕ_i,Σ_i) defines a set function $m : \Delta = \bigcup \Sigma_i \longrightarrow E$ by $m(D) = \int_D \phi_j\, dP$ ($D \in \Sigma_i$, $i \leq j$) whose average range $\{m(D)/P(D) : D\in\Delta, P(D) > 0\}$ is contained in C if $\phi_i \in L^o(\Omega,\Sigma,P,C)$ for all i and if C is measure convex.

Remarks. If C satisfies one of the conditions ① , ② , ③ then also every closed subset of C does. The conditions ② and ③ are obviously satisfied if C is compact since then the least upper bound exists by the Riesz representation theorem. Hence the implication ② ⇒ ④ may be considered as a generalization of the well known Radon-Nikodym theorems for vector measures with relatively compact average range. Note that we did not use one of these theorems in our proof(even though Cor. 1.7 is almost of this form).

We mention without proof some further results from |10|: 1. If C is measure convex then the conditions of the theorem are also equivalent to the following

②' Every increasing chain in $(\mathcal{P}_t(C),\prec)$ has an upper bound in $\mathcal{P}_t(C)$.

If C is not measure convex then the following condition is sufficient for the existence of maximal representing measures for all elements of C

② '' Every increasing chain in $(\mathcal{P}_{t,r}(C),\prec)$ converges to its least upper bound.

Condition ②'' can also be formulated by convergence properties of martingales. It is sometimes more easy to verify than ②. For any completely regular space T the set $C = \mathcal{P}_t(T)$ satisfies ②'' and ④ but not necessarily ②. If C in Theorem 2.2 is analytic (Suslin) then ② and ②'' are equivalent.

Recall that for a (not necessarily convex) set C the point $x \in C$ is called an <u>extreme point</u> if x is not in the convex hull of $C\setminus\{x\}$. If C is analytic then the set ex C is universially measurable and every $\prec$-maximal measure $\mu \in \mathcal{P}_t(C)$ satisfies $\mu(\text{ex } C) = 1$.(This can be proved essentially in the same way as Edgar does it in the convex case in |5|). Thus the following nonconvex version of 'Edgar's theorem' is true:

<u>2.3 Proposition</u>: A Banach space E has the RNP if and only if in every separable closed bounded set $C \subset E$ for each point $x \in C$ there is a measure $\mu \in \mathcal{P}_t(C)$ such that $\mu(\text{ex } C) = 1$ and $r(\mu) = x$.

The 'if'-part follows from the result of Huff and Morris according to which a Banach space E has the Radon-Nikodym-Property iff ex $C \neq \emptyset$ for every separable bounded closed set $C \neq \emptyset$ in E. The analogous question for convex sets still seems to be open.

We conclude by giving an example of a completely regular space T for which $\mathcal{P}_t(T)$ with the topology of pointwise convergence on $C_b(T)$ does not satisfy the conditions ①, ②, ③ of Theorem 2.2. It has been used already by L. Schwartz in order to show that $\mathcal{P}_t(T)$ generally is not measure convex.

Let λ^2 be the two-dimensional Lebesgue measure on $[0,1]^2$ and let λ^2_* denote the corresponding inner measure on the power set of $[0,1]^2$. There is a set $T \subset [0,1]^2$ such that i) $\lambda^2_*(T) = 0$ and ii) card $\{s' \in [0,1] : (s,s') \notin T\} \leq 2$ for each $s \in [0,1]$.

We construct a $\mathcal{P}_t(T)$-valued martingale which does not convrge to a $\mathcal{P}_t(T)$ valued strongly measurable function. We take $\Omega = [0,1]^2$, $\Sigma = \{\lambda^2\text{-measurable sets}\}$ and $P = \lambda^2$ as probability space. The index set is the set I (ordered by inclusion) of the finite partitions of $[0,1]$ into intervals of positive length. For each $i \in I$ Ξ_i is the coresponding σ-algebra over $[0,1]$ and Σ_i is the product σ-algebra $\mathcal{B}([0,1]) \otimes \Xi_i$. We define $\phi_i(s,s')$ for $(s,s') \in [0,1]^2$ and $i \in I$ to be the uniform probability distribution on $T \cap (\{s\} \times J_{i,s'})$ where $J_{i,s'}$ is the interval of the partition i in which the point s' lies. Then it is easy to check that (ϕ_i,Σ_i) is a martingale with values in $\mathcal{P}_t(T)$. For each $i \in I$ and each bounded Σ_i measurable function we have $\int_{[0,1]^2} \langle f|T,\phi_i(s,s')\rangle \, d\lambda^2(s,s') = \int_{[0,1]^2} f(s,s') \, d\lambda^2(s,s')$. So the only possible limit function ϕ must satisfy $\phi(s,s')$ = point measure in (s,s') for almost all (s,s'). But this is not $\mathcal{P}_t(T)$ valued almost everywhere.

Note: At the conference G.A. Edgar kindly gave me a copy of his paper |5| which also appears in this volume. He proves among other related results a version of Theorem 2.2 without using Doob's result. I have changed some notations used in 10 for the convenience of the reader of his paper.

References

1. Choquet, G.: Representations integrales dans les cones convexes sans base compact. C.R. Acad. Sci. Paris 253 (1961) 1901-1903.

2. Doob,J.L.: Generalized sweeping out and probability. J. Funct. An. 2 (1968) 207 - 225.

3. Edgar, G.A.: Extremal Integral Representations, J. Funct. An. 23 (1976) 145-161.

4. -, Measurable weak sections. Illinois J. of Math. 20 (1976) 630-646.

5. On the Radon-Nikodym Property and martingale convergence. In this volume.

6. Ionescu Tulcea,A. and Ionescu Tulcea,C.: Topics in the theory of lifting. Erg. de. Math. 48. Berlin, Heidelberg, New York: Springer 1969.

7. Kupka, J.: Radon-Nikodym theorems for vector valued measures. Trans. AMS 169 (1972) 197-217.

8. Schwartz,L. : Surmartingales regulieres a valeurs mesures et desintegrations reguliers d'une mesure. J. d'an. math. 27 (1973) 1-168.

9. Topsøe, F.: Topology and measure. Lecture Notes in Math. 133. Berlin, Heidelberg, New York: Springer 1970.

10. Weizsäcker, H.v.: Einige maßtheoretische Formen der Sätze von Krein-Milman und Choquet. Habilitationsschrift. München, Februar 1970.

Heinrich von Weizsäcker
Fachbereich Mathematik der Universität
Lahnberge
355 Marburg
West-Germany

8480-78-13
5-42

Vol. 489: J. Bair and R. Fourneau, Etude Géométrique des Espaces Vectoriels. Une Introduction. VII, 185 pages. 1975.

Vol. 490: The Geometry of Metric and Linear Spaces. Proceedings 1974. Edited by L. M. Kelly. X, 244 pages. 1975.

Vol. 491: K. A. Broughan, Invariants for Real-Generated Uniform Topological and Algebraic Categories. X, 197 pages. 1975.

Vol. 492: Infinitary Logic: In Memoriam Carol Karp. Edited by D. W. Kueker. VI, 206 pages. 1975.

Vol. 493: F. W. Kamber and P. Tondeur, Foliated Bundles and Characteristic Classes. XIII, 208 pages. 1975.

Vol. 494: A Cornea and G. Licea. Order and Potential Resolvent Families of Kernels. IV, 154 pages. 1975.

Vol. 495: A. Kerber, Representations of Permutation Groups II. V, 175 pages. 1975.

Vol. 496: L. H. Hodgkin and V. P. Snaith, Topics in K-Theory. Two Independent Contributions. III, 294 pages. 1975.

Vol. 497: Analyse Harmonique sur les Groupes de Lie. Proceedings 1973–75. Edité par P. Eymard et al. VI, 710 pages. 1975.

Vol. 498: Model Theory and Algebra. A Memorial Tribute to Abraham Robinson. Edited by D. H. Saracino and V. B. Weispfenning. X, 463 pages. 1975.

Vol. 499: Logic Conference, Kiel 1974. Proceedings. Edited by G. H. Müller, A. Oberschelp, and K. Potthoff. V, 651 pages 1975.

Vol. 500: Proof Theory Symposion, Kiel 1974. Proceedings. Edited by J. Diller and G. H. Müller. VIII, 383 pages. 1975.

Vol. 501: Spline Functions, Karlsruhe 1975. Proceedings. Edited by K. Böhmer, G. Meinardus, and W. Schempp. VI, 421 pages. 1976.

Vol. 502: János Galambos, Representations of Real Numbers by Infinite Series. VI, 146 pages. 1976.

Vol. 503: Applications of Methods of Functional Analysis to Problems in Mechanics. Proceedings 1975. Edited by P. Germain and B. Nayroles. XIX, 531 pages. 1976.

Vol. 504: S. Lang and H. F. Trotter, Frobenius Distributions in GL_2-Extensions. III, 274 pages. 1976.

Vol. 505: Advances in Complex Function Theory. Proceedings 1973/74. Edited by W. E. Kirwan and L. Zalcman. VIII, 203 pages. 1976.

Vol. 506: Numerical Analysis, Dundee 1975. Proceedings. Edited by G. A. Watson. X, 201 pages. 1976.

Vol. 507: M. C. Reed, Abstract Non-Linear Wave Equations. VI, 128 pages. 1976.

Vol. 508: E. Seneta, Regularly Varying Functions. V, 112 pages. 1976.

Vol. 509: D. E. Blair, Contact Manifolds in Riemannian Geometry. VI, 146 pages. 1976.

Vol. 510: V. Poènaru, Singularités C^∞ en Présence de Symétrie. V, 174 pages. 1976.

Vol. 511: Séminaire de Probabilités X. Proceedings 1974/75. Edité par P. A. Meyer. VI, 593 pages. 1976.

Vol. 512: Spaces of Analytic Functions, Kristiansand, Norway 1975. Proceedings. Edited by O. B. Bekken, B. K. Øksendal, and A. Stray. VIII, 204 pages. 1976.

Vol. 513: R. B. Warfield, Jr. Nilpotent Groups. VIII, 115 pages. 1976.

Vol. 514: Séminaire Bourbaki vol. 1974/75. Exposés 453 – 470. IV, 276 pages. 1976.

Vol. 515: Bäcklund Transformations. Nashville, Tennessee 1974. Proceedings. Edited by R. M. Miura. VIII, 295 pages. 1976.

Vol. 516: M. L. Silverstein, Boundary Theory for Symmetric Markov Processes. XVI, 314 pages. 1976.

Vol. 517: S. Glasner, Proximal Flows. VIII, 153 pages. 1976.

Vol. 518: Séminaire de Théorie du Potentiel, Proceedings Paris 1972–1974. Edité par F. Hirsch et G. Mokobodzki. VI, 275 pages. 1976.

Vol. 519: J. Schmets, Espaces de Fonctions Continues. XII, 150 pages. 1976.

Vol. 520: R. H. Farrell, Techniques of Multivariate Calculation. X, 337 pages. 1976.

Vol. 521: G. Cherlin, Model Theoretic Algebra – Selected Topics. IV, 234 pages. 1976.

Vol. 522: C. O. Bloom and N. D. Kazarinoff, Short Wave Radiation Problems in Inhomogeneous Media: Asymptotic Solutions. V. 104 pages. 1976.

Vol. 523: S. A. Albeverio and R. J. Høegh-Krohn, Mathematical Theory of Feynman Path Integrals. IV, 139 pages. 1976.

Vol. 524: Séminaire Pierre Lelong (Analyse) Année 1974/75. Edité par P. Lelong. V, 222 pages. 1976.

Vol. 525: Structural Stability, the Theory of Catastrophes, and Applications in the Sciences. Proceedings 1975. Edited by P. Hilton. VI, 408 pages. 1976.

Vol. 526: Probability in Banach Spaces. Proceedings 1975. Edited by A. Beck. VI, 290 pages. 1976.

Vol. 527: M. Denker, Ch. Grillenberger, and K. Sigmund, Ergodic Theory on Compact Spaces. IV, 360 pages. 1976.

Vol. 528: J. E. Humphreys, Ordinary and Modular Representations of Chevalley Groups. III, 127 pages. 1976.

Vol. 529: J. Grandell, Doubly Stochastic Poisson Processes. X, 234 pages. 1976.

Vol. 530: S. S. Gelbart, Weil's Representation and the Spectrum of the Metaplectic Group. VII, 140 pages. 1976.

Vol. 531: Y.-C. Wong, The Topology of Uniform Convergence on Order-Bounded Sets. VI, 163 pages. 1976.

Vol. 532: Théorie Ergodique. Proceedings 1973/1974. Edité par J.-P. Conze and M. S. Keane. VIII, 227 pages. 1976.

Vol. 533: F. R. Cohen, T. J. Lada, and J. P. May, The Homology of Iterated Loop Spaces. IX, 490 pages. 1976.

Vol. 534: C. Preston, Random Fields. V, 200 pages. 1976.

Vol. 535: Singularités d'Applications Differentiables. Plans-sur-Bex. 1975. Edité par O. Burlet et F. Ronga. V, 253 pages. 1976.

Vol. 536: W. M. Schmidt, Equations over Finite Fields. An Elementary Approach. IX, 267 pages. 1976.

Vol. 537: Set Theory and Hierarchy Theory. Bierutowice, Poland 1975. A Memorial Tribute to Andrzej Mostowski. Edited by W. Marek, M. Srebrny and A. Zarach. XIII, 345 pages. 1976.

Vol. 538: G. Fischer, Complex Analytic Geometry. VII, 201 pages. 1976.

Vol. 539: A. Badrikian, J. F. C. Kingman et J. Kuelbs, Ecole d'Eté de Probabilités de Saint Flour V-1975. Edité par P.-L. Hennequin. IX, 314 pages. 1976.

Vol. 540: Categorical Topology, Proceedings 1975. Edited by E. Binz and H. Herrlich. XV, 719 pages. 1976.

Vol. 541: Measure Theory, Oberwolfach 1975. Proceedings. Edited by A. Bellow and D. Kölzow. XIV, 430 pages. 1976.

Vol. 542: D. A. Edwards and H. M. Hastings, Čech and Steenrod Homotopy Theories with Applications to Geometric Topology. VII, 296 pages. 1976.

Vol. 543: Nonlinear Operators and the Calculus of Variations, Bruxelles 1975. Edited by J. P. Gossez, E. J. Lami Dozo, J. Mawhin, and L. Waelbroeck, VII, 237 pages. 1976.

Vol. 544: Robert P. Langlands, On the Functional Equations Satisfied by Eisenstein Series. VII, 337 pages. 1976.

Vol. 545: Noncommutative Ring Theory. Kent State 1975. Edited by J. H. Cozzens and F. L. Sandomierski. V, 212 pages. 1976.

Vol. 546: K. Mahler, Lectures on Transcendental Numbers. Edited and Completed by B. Diviš and W. J. Le Veque. XXI, 254 pages. 1976.

Vol. 547: A. Mukherjea and N. A. Tserpes, Measures on Topological Semigroups: Convolution Products and Random Walks. V, 197 pages. 1976.

Vol. 548: D. A. Hejhal, The Selberg Trace Formula for PSL $(2, \mathbb{R})$. Volume I. VI, 516 pages. 1976.

Vol. 549: Brauer Groups, Evanston 1975. Proceedings. Edited by D. Zelinsky. V, 187 pages. 1976.

Vol. 550: Proceedings of the Third Japan – USSR Symposium on Probability Theory. Edited by G. Maruyama and J. V. Prokhorov. VI, 722 pages. 1976.

Vol. 551: Algebraic K-Theory, Evanston 1976. Proceedings. Edited by M. R. Stein. XI, 409 pages. 1976.

Vol. 552: C. G. Gibson, K. Wirthmüller, A. A. du Plessis and E. J. N. Looijenga. Topological Stability of Smooth Mappings. V, 155 pages. 1976.

Vol. 553: M. Petrich, Categories of Algebraic Systems. Vector and Projective Spaces, Semigroups, Rings and Lattices. VIII, 217 pages. 1976.

Vol. 554: J. D. H. Smith, Mal'cev Varieties. VIII, 158 pages. 1976.

Vol. 555: M. Ishida, The Genus Fields of Algebraic Number Fields. VII, 116 pages. 1976.

Vol. 556: Approximation Theory. Bonn 1976. Proceedings. Edited by R. Schaback and K. Scherer. VII, 466 pages. 1976.

Vol. 557: W. Iberkleid and T. Petrie, Smooth S^1 Manifolds. III, 163 pages. 1976.

Vol. 558: B. Weisfeiler, On Construction and Identification of Graphs. XIV, 237 pages. 1976.

Vol. 559: J.-P. Caubet, Le Mouvement Brownien Relativiste. IX, 212 pages. 1976.

Vol. 560: Combinatorial Mathematics, IV, Proceedings 1975. Edited by L. R. A. Casse and W. D. Wallis. VII, 249 pages. 1976.

Vol. 561: Function Theoretic Methods for Partial Differential Equations. Darmstadt 1976. Proceedings. Edited by V. E. Meister, N. Weck and W. L. Wendland. XVIII, 520 pages. 1976.

Vol. 562: R. W. Goodman, Nilpotent Lie Groups: Structure and Applications to Analysis. X, 210 pages. 1976.

Vol. 563: Séminaire de Théorie du Potentiel. Paris, No. 2. Proceedings 1975–1976. Edited by F. Hirsch and G. Mokobodzki. VI, 292 pages. 1976.

Vol. 564: Ordinary and Partial Differential Equations, Dundee 1976. Proceedings. Edited by W. N. Everitt and B. D. Sleeman. XVIII, 551 pages. 1976.

Vol. 565: Turbulence and Navier Stokes Equations. Proceedings 1975. Edited by R. Temam. IX, 194 pages. 1976.

Vol. 566: Empirical Distributions and Processes. Oberwolfach 1976. Proceedings. Edited by P. Gaenssler and P. Révész. VII, 146 pages. 1976.

Vol. 567: Séminaire Bourbaki vol. 1975/76. Exposés 471–488. IV, 303 pages. 1977.

Vol. 568: R. E. Gaines and J. L. Mawhin, Coincidence Degree, and Nonlinear Differential Equations. V, 262 pages. 1977.

Vol. 569: Cohomologie Etale SGA 4½. Séminaire de Géométrie Algébrique du Bois-Marie. Edité par P. Deligne. V, 312 pages. 1977.

Vol. 570: Differential Geometrical Methods in Mathematical Physics, Bonn 1975. Proceedings. Edited by K. Bleuler and A. Reetz. VIII, 576 pages. 1977.

Vol. 571: Constructive Theory of Functions of Several Variables, Oberwolfach 1976. Proceedings. Edited by W. Schempp and K. Zeller. VI. 290 pages. 1977

Vol. 572: Sparse Matrix Techniques, Copenhagen 1976. Edited by V. A. Barker. V, 184 pages. 1977.

Vol. 573: Group Theory, Canberra 1975. Proceedings. Edited by R. A. Bryce, J. Cossey and M. F. Newman. VII, 146 pages. 1977.

Vol. 574: J. Moldestad, Computations in Higher Types. IV, 203 pages. 1977.

Vol. 575: K-Theory and Operator Algebras, Athens, Georgia 1975. Edited by B. B. Morrel and I. M. Singer. VI, 191 pages. 1977.

Vol. 576: V. S. Varadarajan, Harmonic Analysis on Real Reductive Groups. VI, 521 pages. 1977.

Vol. 577: J. P. May, E_∞ Ring Spaces and E_∞ Ring Spectra. IV, 268 pages. 1977.

Vol. 578: Séminaire Pierre Lelong (Analyse) Année 1975/76. Edité par P. Lelong. VI, 327 pages. 1977.

Vol. 579: Combinatoire et Représentation du Groupe Symétrique, Strasbourg 1976. Proceedings 1976. Edité par D. Foata. IV, 339 pages. 1977.

Vol. 580: C. Castaing and M. Valadier, Convex Analysis and Measurable Multifunctions. VIII, 278 pages. 1977.

Vol. 581: Séminaire de Probabilités XI, Université de Strasbourg. Proceedings 1975/1976. Edité par C. Dellacherie, P. A. Meyer et M. Weil. VI, 574 pages. 1977.

Vol. 582: J. M. G. Fell, Induced Representations and Banach *-Algebraic Bundles. IV, 349 pages. 1977.

Vol. 583: W. Hirsch, C. C. Pugh and M. Shub, Invariant Manifolds. IV, 149 pages. 1977.

Vol. 584: C. Brezinski, Accélération de la Convergence en Analyse Numérique. IV, 313 pages. 1977.

Vol. 585: T. A. Springer, Invariant Theory. VI, 112 pages. 1977.

Vol. 586: Séminaire d'Algèbre Paul Dubreil, Paris 1975–1976 (29ème Année). Edited by M. P. Malliavin. VI, 188 pages. 1977.

Vol. 587: Non-Commutative Harmonic Analysis. Proceedings 1976. Edited by J. Carmona and M. Vergne. IV, 240 pages. 1977.

Vol. 588: P. Molino, Théorie des G-Structures: Le Problème d'Equivalence. VI, 163 pages. 1977.

Vol. 589: Cohomologie l-adique et Fonctions L. Séminaire de Géométrie Algébrique du Bois-Marie 1965–66, SGA 5. Edité par L. Illusie. XII, 484 pages. 1977.

Vol. 590: H. Matsumoto, Analyse Harmonique dans les Systèmes de Tits Bornologiques de Type Affine. IV, 219 pages. 1977.

Vol. 591: G. A. Anderson, Surgery with Coefficients. VIII, 157 pages. 1977.

Vol. 592: D. Voigt, Induzierte Darstellungen in der Theorie der endlichen, algebraischen Gruppen. V, 413 Seiten. 1977.

Vol. 593: K. Barbey and H. König, Abstract Analytic Function Theory and Hardy Algebras. VIII, 260 pages. 1977.

Vol. 594: Singular Perturbations and Boundary Layer Theory, Lyon 1976. Edited by C. M. Brauner, B. Gay, and J. Mathieu. VIII, 539 pages. 1977.

Vol. 595: W. Hazod, Stetige Faltungshalbgruppen von Wahrscheinlichkeitsmaßen und erzeugende Distributionen. XIII, 157 Seiten. 1977.

Vol. 596: K. Deimling, Ordinary Differential Equations in Banach Spaces. VI, 137 pages. 1977.

Vol. 597: Geometry and Topology, Rio de Janeiro, July 1976. Proceedings. Edited by J. Palis and M. do Carmo. VI, 866 pages. 1977.

Vol. 598: J. Hoffmann-Jørgensen, T. M. Liggett et J. Neveu, Ecole d'Eté de Probabilités de Saint-Flour VI – 1976. Edité par P.-L. Hennequin. XII, 447 pages. 1977.

Vol. 599: Complex Analysis, Kentucky 1976. Proceedings. Edited by J. D. Buckholtz and T. J. Suffridge. X, 159 pages. 1977.

Vol. 600: W. Stoll, Value Distribution on Parabolic Spaces. VIII, 216 pages. 1977.

Vol. 601: Modular Functions of one Variable V, Bonn 1976. Proceedings. Edited by J.-P. Serre and D. B. Zagier. VI, 294 pages. 1977.

Vol. 602: J. P. Brezin, Harmonic Analysis on Compact Solvmanifolds. VIII, 179 pages. 1977.

Vol. 603: B. Moishezon, Complex Surfaces and Connected Sums of Complex Projective Planes. IV, 234 pages. 1977.

Vol. 604: Banach Spaces of Analytic Functions, Kent, Ohio 1976. Proceedings. Edited by J. Baker, C. Cleaver and Joseph Diestel. VI, 141 pages. 1977.

Vol. 605: Sario et al., Classification Theory of Riemannian Manifolds. XX, 498 pages. 1977.

Vol. 606: Mathematical Aspects of Finite Element Methods. Proceedings 1975. Edited by I. Galligani and E. Magenes. VI, 362 pages. 1977.

Vol. 607: M. Métivier, Reelle und Vektorwertige Quasimartingale und die Theorie der Stochastischen Integration. X, 310 Seiten. 1977.

Vol. 608: Bigard et al., Groupes et Anneaux Réticulés. XIV, 334 pages. 1977.